Time Reversal Acoustics

Woon Siong Gan

Time Reversal Acoustics

Springer

Woon Siong Gan
Acoustical Technologies Singapore Pte Ltd
Singapore, Singapore

ISBN 978-981-16-3234-1 ISBN 978-981-16-3235-8 (eBook)
https://doi.org/10.1007/978-981-16-3235-8

This Springer imprint is published by the registered company Springer Nature Singapore Pte Ltd.
The registered company address is: 152 Beach Road, #21-01/04 Gateway East, Singapore 189721, Singapore

To My Parents

To My Family

Foreword

Time reversal is an ancient obsession of physicists. One of the most intriguing symmetries in physics is the time-reversal invariance. It is a very rich and powerful concept both in classical and in quantum mechanics. It has an impact on all facets of physics and important applications in quantum field theory and particle physics. However, simple experimental evidence of this concept is difficult to obtain in the field of classical mechanics. Although the equations of classical mechanics are time-reversal invariant on the microscopic scale, it is difficult to test this invariance in a complex system of particles. Such a test requires a time-reversal mirror, and it has been considered only as 'a thought experiments'. However in the field of acoustic wave propagation, where time-reversal invariance also occurs in many situations, such a time-reversal mirror can be built easily. It is in the field of acoustics that the first time-reversal mirrors were realized in the nineties with today electronics, and thanks to large array of reversible acousto-electric transducers, opening the field of time-reversal acoustics.

Professor Woon Siong Gan offers in this book on "time-reversal acoustics" both an interesting approach and a detailed review of this field, with both theoretical and experimental developments. One of the original features of his presentation is to introduce time-reversal symmetry as one of the various symmetries that can exists in physics. He seeks to understand how various symmetry concepts govern acoustic wave propagation in various media.

Beginning with the fundamental theorem on symmetry in physics, the Noether's theorem, he first extends to acoustics the concept of Gauge invariance originated initially from electromagnetics. Then, he explored the notion of Lie group to model the various symmetries of partial differential equation as the one describing acoustic wave propagation.

After a presentation of this formalism, he introduced the concept of time-reversal mirrors in linear acoustics and he extends the discussion by introducing the acoustic sink approach as related to time-reversed source.

How nonlinear acoustics can break reversibility is discussed as well as the effect of non-reciprocal acoustic medium? Following these chapters, he introduced the principle of microscopic reversibility, as the root of time-reversal acoustics.

Professor Woon Siong Gan extends this approach to various symmetry laws that govern the properties of material, and he discusses the way waves propagate in media governed by parity-time invariance. Then, he introduced non-reciprocal medium and discussed topological sound wave propagation.

A comparison with quantum time-reversal symmetry and the way to break this symmetry is also presented with the description of various interesting effects, for instance the "acoustical Quantum Hall effect".

Then, in the last chapters of this book, he describes various applications of time-reversal acoustics in signal processing, underwater communications, sound control in complex medium, seismology, biomedical ultrasound imaging and therapy and non-destructive testing.

Hopefully, this book will attract many researchers and students to the study of wave equation symmetries and to the various applications of these concepts to create new materials for wave control in complex media.

May 2021

Mathias Fink
ESPCI Paris
Paris, France

Preface

The unique feature of this book is the in-depth treatment of the theory of time reversal acoustics with first eleven chapters dedicated to the fundamentals followed by the remaining four chapters on the applications. Time reversal acoustics is an exemplification of the symmetry properties of acoustic fields and an application of the gauge invariance approach to acoustic fields. In 2007, Woon Siong Gan introduced gauge theory into acoustics or the gauge invariance approach to acoustic fields. The scope of time reversal acoustics has been extended from the original linear and classical regime to the nonlinear and quantum regime. First, the mathematical tool of Lie group and Lie algebra is introduced. Then nonreciprocal acoustics and Onsager reciprocal relations are described. The root of time reversal acoustics is traced to the principle of microscopic reversibility. Parity time invariance as a fuller aspect of time reversal acoustics is given which is demonstrated in quantum time reversal acoustics and time reversal acoustics in the complex plane. Spontaneous time reversal symmetry breaking is given as part of the time arrow problem. The theoretical portion of this book is rounded up with the topics on the topology in acoustics and topological sound waves with macroscopic structure demonstrating quantum effects of acoustic quantum Hall effect, acoustical quantum spin Hall effect and acoustical valley Hall effect. The application portion of the book consists of four chapters on the applications of time reversal acoustics to signal processing and underwater communications, seismic exploration, nondestructive testing and medical ultrasound imaging.

Singapore
May 2021

Woon Siong Gan

Contents

Chapter 1
Symmetry Properties of Acoustic Fields

1.1 Gauge Invariance Approach to Sound Propagation in Solids

In 2007, Gan [1] first introduced symmetry properties to acoustic fields and gauge theory to acoustics. Examples of symmetry properties of acoustic fields are time-reversal symmetry of acoustic fields, form invariance of acoustic field equations and the broken spatial translational symmetry giving rise to phonon which is a Goldstone mode. Group theory is the mathematics of symmetry. The symmetry properties of acoustic fields when propagating through a solid medium will normally undergo translational symmetry and rotational symmetry. The time-reversal symmetry is another symmetry property of the acoustic field in addition to the above two forms of symmetry. To understand the symmetry properties of the acoustic fields, one needs to briefly introduce the following topics: global gauge invariance, local gauge invariance, covariant derivative, Lie group, spontaneous symmetry breaking and Noether's theorem [2].

Gauge invariance originated from electromagnetic wavefields. The Maxwell's equations are the first gauge theory. Here, the structure of the electromagnetic field equations is such that the electric field and the magnetic field are expressed in terms of the gauge fields, the scalar potential, and the vector potential. Gauge invariance means that that this whole class of scalar and vector potentials related by gauge transformation resulting in no change in the form of the equations. This concept of gauge invariance for Maxwell's equations has been extended by Yang Mills theory [3] to a more general theory. It has also been extended by Gan [1] in 2007 to the acoustic wavefields and the acoustic equation of motion. Here, the acoustic wavefields, the stress field, and the particle velocity field are expressed in terms of the gauge fields, the acoustic scalar potential, and the acoustic vector potential with gauge invariance in the acoustic equation of motion. The concept of gauge invariance is further developed into the classification of global gauge invariance and local gauge invariance with the introduction of the Lagrangian. The Lagrangian is invariant under the gauge transformation. Local gauge invariance is different from global gauge invariance

W. S. Gan, *Time Reversal Acoustics*,
https://doi.org/10.1007/978-981-16-3235-8_1

with the introduction of the interaction Lagrangian. Thus, local gauge invariance is more stringent than the global gauge invariance. Weyl's [4] gauge principle states that when the original Lagrangian is not locally gauge invariant, a new gauge field must be introduced in order to satisfy local gauge invariance. And the Lagrangian is then to be altered by replacing the partial derivative with the covariant derivative. Noether's theorem [2] is an important theorem in gauge theory which states that for every symmetry, there is a corresponding conservation law. Noether's theorem [2] does not work for discrete symmetry like time-reversal symmetry. Instead energy conservation follows from time translation symmetry which is a continuous symmetry. The time-reversal symmetry's mathematics is under the Poincare group which is ten dimensional non-Abelian Lie group. As a topological space, the group has four connected components: the component of the identity, the time-reversal component, the spatial inversion component, and the component which is both time-reversed and spatially inverted. To have a deeper understanding of time-reversal symmetry, one need to go through the Poincare group which covers space inversion and time reversal.

For sound propagation in solids, one has to consider a moving medium instead of a stationary medium and the unstressed state of the material is evolving with time. The type of gauge transformation or gauge symmetry applicable to sound propagation in solids is Galilean transformation or Galilean symmetry which has to include both rotational and translational symmetry. A gauge theory for ideal fluid flow based on Galilean transformations and covariant derivative was derived by Kambe [5]. These are properties of the gauge transformation intrinsic to the acoustic equation of motion. In this section, one extends the gauge invariance formulation to sound propagation in solids. There is the analogy for the electromagnetic fields that the covariant derivative is also intrinsic to the Maxwell's equations. The different nature between electromagnetic waves and sound waves gives rise to the covariant derivative for the Maxwell's equations leads to the Lorentz's transformation and the covariant derivative for the acoustic equation of motion leads to the Galilean transformation. When the velocity of the medium is much less than the velocity of light, the Lorentz transformation reduces to the Galilean transformation.

The introduction of the covariant derivative is to fulfil the requirement of local gauge invariance, and it also satisfies the Galilean transformation. This can be shown as:

$$D_t := \partial_t + G \tag{1.1}$$

where $D_t =$ covariant derivative, and $G =$ new gauge field.

The Galilean transformation will be used to describe sound propagation in solids. A Galilean transformation in Newtonian mechanics is the translational transformation from one coordinate system A to coordinate system A' moving with a relative velocity R. The transformation law is given by:

$$X = (t, x) \rightarrow x' = \left(t', x'\right) = (t, x - Rt) \tag{1.2}$$

This is a sequence of global translational transformations with parameter t.

To extend this to local gauge invariance, Kambe [5] has derived the covariant derivative given by:

$$D_t = \partial_t + (v \cdot \nabla) \tag{1.3}$$

where v = phase velocity.

The following will give two examples of the use of covariant derivative in the gauge theory formulation of sound propagation in solids.

1.1.1 Infinitesimal Amplitude Sound Wave Propagation

An acoustic wavefield equation derived from Newton's law of motion can be given by:

$$\nabla \cdot \vec{T} = \rho \frac{\partial^2}{\partial t^2} \vec{u} - \vec{F} \tag{1.4}$$

where $\vec{T}$ = stress, $\vec{u}$ = displacement, $\vec{F}$ = body force, and ρ = density of the medium.

Replacing the partial derivative in (1.4) by Eq. (1.3), one obtains:

$$\nabla \cdot \vec{\boldsymbol{T}} = \rho \left(\frac{\partial \vec{\boldsymbol{v}}}{\partial t} + (\vec{\boldsymbol{v}} \cdot \nabla) \vec{\boldsymbol{v}} \right) - \vec{\boldsymbol{F}} \tag{1.5}$$

If sound propagation only in the x-direction, and with $\vec{F} = 0$ for source-free region, (1.5) can be reduced to a simpler form shown below:

$$\frac{\partial^2 \vec{u}}{\partial x'^2} = \rho \frac{\partial^2 \vec{u}}{\partial t^2} + \rho \frac{\partial \vec{u}}{\partial t} \frac{\partial^2 \vec{u}}{\partial x' \partial t} \tag{1.6}$$

where x' = moving coordinate given by $x' = x - Rt$.

It shows that with the introduction of the covariant derivative, there is the additional second term on the right-hand side of the equation. So far, no one has attempted on the exact analytical solution of this equation yet.

1.1.2 Finite Amplitude Sound Wave Propagation

When applying the covariant derivative (1.3) to the nonlinear wave equation given by (1.7),

$$\ddot{\vec{u}} = \frac{M_2}{\rho}\frac{\partial^2 \vec{u}}{\partial x^2}\left[1 + \frac{M_3}{M_2}\frac{\partial \vec{u}}{\partial x}\right] \tag{1.7}$$

where $M_2 = K_2$, $M_3 = K_3 + 2K_2$, K_2, and K_3 are related to the second- and third-order elastic constants.

The following equation is obtained:

$$M_2\frac{\partial^2 \vec{u}}{\partial x'^2}\left[1 + \frac{M_3}{M_2}\frac{\partial \vec{u}}{\partial x'}\right] = \rho\left(\frac{\partial^2 \vec{u}}{\partial t^2} + \frac{\partial \vec{u}}{\partial t}\frac{\partial^2 \vec{u}}{\partial x'\partial t}\right) \tag{1.8}$$

The introduction of the covariant derivative only introduces the same additional term on the right-hand side of the equation as in the case for the linear wave equation in (1.6). Again so far, no one has obtained the exact analytical solution of this equation yet.

1.2 Components of Time-Reversal Symmetry

There are two components of the time reversal. The first is the order in time reversal. If the changing states of a physical system is represented by a curve S: $R \to P$ through some set of states P with initial state $S(0)\ \varepsilon\ P$, then the time reversal will turn around the temporal order in which such states occur. Thus, time-reversal transformation transforms the temporal parameter t as: $t \to -t$. The second component of time reversal is the time-reversal operator.

1.3 The Poincare Group and Abelian Group Properties

Most theories in physics can be represented by a state space P and a set S of preferred possible trajectories which are the solutions to some law. The time invariance property of time-reversal symmetry can be illustrated by the example of the Newtonian mechanics. Here, a point particle of mass m in the presence of a force F, the states are the possible positions of the particle in space and the trajectories are the curves with acceleration a satisfies Newton's second law: $F = m\frac{\mathrm{d}^2}{\mathrm{d}t^2}x = ma$. Consider a set of curves through a state space P and let S be the subset of the curves through P which are solutions to some law. Take a bijection ϕ on the set of curves through P to be an invariance or a symmetry of the law means that if some curve is a solution to the law, then the ϕ-transformed curve is also a solution. This can be put in this way: if a law has a solution set S, then ϕ is a symmetry of the law if and only if $\phi(S) = S$. Time-reversal invariance is only a special case of this. That is, for a law to be time reversal invariant means that a curve is a solution to the law only if the time-reversal curve is a solution as well. This can be illustrated by if $x(t)$ is a solution to Newton's equation, then so is its time reverse $x(-t)$.

1.3.1 Properties of Poincare Group

The definition of the Poincare group is as follows: the Poincare group $P(1, n)$ in n space dimension is the set of matrices

$$R(L,b) = \begin{bmatrix} L_{00} & \cdots & L_{0n} & b_0 \\ \vdots & & \vdots & \vdots \\ L_{n0} & & L_{nn} & b_n \\ 0 & & 0 & 1 \end{bmatrix} \tag{1.9}$$

where $L \in 0(1, n)$ is a Lorentz transformation, that is $L^{\mathrm{T}}gL = g$ with $g = \mathrm{diag}(1, -1, \ldots, -1)$ is a diagonal matrix of rank $(n + 1)$ and b the $(n + 1)$ component column vector. G is of rank $n + 2$. Its action on a vector ξ with transpose $\xi^{\mathrm{T}} = (x_0, x_1, \ldots, x_n, 1)$ is represented by

$$R(L,b)\xi = \begin{bmatrix} L_{00} & \cdots & L_{0n} & b_0 \\ \vdots & & \vdots & \vdots \\ L_{n0} & & L_{nn} & b_n \\ 0 & & 0 & 1 \end{bmatrix} \begin{pmatrix} x_0 \\ \vdots \\ x_n \\ 1 \end{pmatrix} = \begin{pmatrix} x'_0 \\ \vdots \\ x'_n \\ 1 \end{pmatrix} = \xi' \tag{1.10}$$

The time reversal and the space inversion operator T and I can be expressed in terms of elements of $0(1, n)$ as $T \begin{bmatrix} -1 & 0_{1xn} \\ 0_{nx1} & 1_{nxn} \end{bmatrix}$ and $I = \begin{bmatrix} 1 & 0_{1xn} \\ 0_{nx1} & -1_{nxn} \end{bmatrix}$ and both are of rank $(n + 1)$.

The corresponding rank matrices $\mathcal{J}$ and τ and elements of $P(1, n)$ and therefore of rank $(n + 2)$ given by

$$\mathcal{J} = \begin{bmatrix} I_{(n+1)} & 0_{(n+1)x1} \\ 0_{1x(n+1)} & 1 \end{bmatrix}, \quad \tau = \begin{bmatrix} -1 & 0_{1xn} \\ 0_{nx1} & 1_{nxn} \end{bmatrix} \tag{1.11}$$

It is to be noted that $\mathcal{J}^{-1} = \mathcal{J}$ and $\tau^{-1} = \tau$.

The Poincare group can be represented in terms of the Lie algebra as follows. We take the example of the case of the Poincare group consisting of 5×5 matrices. The generator of the Poincare Lie algebra $P(1, 3)$ is classified as translation, rotation, or both.

References

1. Gan, W.S.: Gauge invariance approach to acoustic fields. In: Akiyama, I. (ed.) Acoustical Imaging, vol. 29, pp. 389–394. Springer (2007)

2. Noether, E.: Invariante Variationsprobleme. Nachrichten von der Gesellschaft der Wissenschaften zu Göttingen. Math.-Phys. Klass. 235–257 (1918)
3. Yang, C.N., Mills, R.L.: Conservation of isotopic spin and isotopic gauge invariance. Phys. Rev. **96**(1), 191 (1954)
4. Weyl, H.: Gravitation und Elektrizitat. Sitzungsber. Akademie der Wissenschaften, Berlin, pp. 465–480. Siehe auch die Gesammelten Abhandlungen, 6 Vols. Chadrasekharan, K. (ed.). Springer-Verlag (1918)
5. Kambe, T.: Variational formulation of ideal fluid flows according to gauge principle. Preprint accepted by Fluid Dyn. Res.

Chapter 2
Lie Group and Lie Algebra

2.1 Lie Group [1]

Group theory is the mathematics of symmetry. To understand in depth in symmetry, one has to be familiar with group theory. Lie group is the form of group theory used for symmetry. Lie group was introduced by Sophus Lie in 1873 to model the continuous symmetry of differential equation.

Lie group is a continuous group and thus provides a natural model for the concept of continuous symmetry such as rotational symmetry in three dimensions given by the special orthogonal group. Continuous group is in oppose to discrete groups where the elements are separated. Lie group laid the foundation of the theory of continuous transformation groups.

A key idea in the theory of Lie groups is to replace the global object, the group with its local version, called by Lie himself as its infinitesimal group. This infinitesimal group is since became known as its Lie algebra. An example of Lie group is the set of complex numbers with absolute value 1. This corresponds to the points on the circle of centre 0 and radius 1 in the complex plane. This is a Lie group known as the circle group under complex multiplication.

Lie groups and the associated Lie algebra play an important role in modern physics. They play the role of the symmetry of a physical system. Examples of Lie groups of particular importance are the Poincare group, the rotation group in three dimension SO(3) or special orthogonal group and the special unitary group SU(3).

A real Lie group is a group that is a finite-dimensional real smooth manifold. Take a four-dimensional non-compact real Lie group such as

$$\mathrm{GL}(2, R) = \left\{ A = \begin{pmatrix} a & b \\ c & d \end{pmatrix} : \det A = ad - bc \neq 0 \right\} \tag{2.1}$$

This group is disconnected. It has two connected components corresponding to the positive and negative values of the determinants. The rotation matrices, being a

W. S. Gan, *Time Reversal Acoustics*,
https://doi.org/10.1007/978-981-16-3235-8_2

subgroup of GL(2, R) and denoted by SO(2, R). It is a Lie group and a one-dimensional compact connected Lie group which is diffeomorphic to the circle. With the rotation angle ϕ as a parameter, it can be parameterized as:

$$\mathrm{SO}(2, R) = \left\{ \begin{matrix} \cos\phi & -\sin\phi \\ \sin\phi & \cos\phi \end{matrix} \right\} : \phi \epsilon R/2\pi Z \tag{2.2}$$

Lie group can be studied using differential calculus because they are smooth differentiable manifolds.

2.2 The Orthogonal Group

The orthogonal group is a Lie group and an algebraic group. In n dimension denoted by (n) is group transformation of a Euclidean space of dimension n that preserves a fixed point with group operation given by composing transformations. It is a group consisting of $n \times n$ orthogonal matrices and the group operation is given by matrix multiplication. An orthogonal matrix is a real matrix with inverse equals its transpose. The orthogonal group of dimensions n has two connected components. The one containing the identity element is a subgroup, known as the special orthogonal group, denoted by SO(n). It consists of all orthogonal matrices of determinant 1. This group is also called the rotation group. This generalizes that in dimension 2, the elements are the usual rotation around a point and in dimension 3, and its elements are the usual rotations around a line. Another connected component of the orthogonal group consists of all orthogonal matrices of determinant -1.

The low dimensions of the special orthogonal group such as SO(2), SO(3), and SO(4) have been widely studied. Considering any field F, an orthogonal ($n \times n$) matrix with entries in F will have inverse equals its transpose. This ($n \times n$) orthogonal matrices form a subgroup, denoted by $O(n, F)$ of the general linear group GL(n, F) or

$$O(n, F) = Q\epsilon \mathrm{GL}(n, F)/Q^{\mathrm{T}}Q = QQ^{\mathrm{T}} = I \tag{2.3}$$

Orthogonal groups are algebraic groups because the condition of preserving a form can be expressed as an equation of matrices.

2.2.1 *The SO(*n*) Group*

A common example of an orthogonal group is the SO(n) group with the Euclidean vector space as the orthonormal basis; the orthogonal group can be identified with the group of orthogonal matrices. The orthogonal matrices are defined as matrices fulfilling

$$QQ^{\mathrm{T}} = I \tag{2.4}$$

where I = identity matrix. The identity matrix is sometimes called a unit matrix of size n, is the $n \times n$ square matrix with ones on the main diagonal and zeros elsewhere. It is denoted by I_n. Examples are:

$$I_1 = [1], \quad I_2 = \begin{bmatrix} 1 & 0 \\ 0 & 1 \end{bmatrix}, \quad I_3 = \begin{bmatrix} 1 & 0 & 0 \\ 0 & 1 & 0 \\ 0 & 0 & 1 \end{bmatrix} \text{ etc.}$$

This follows that the square of the determinant of Q equals 1. So the determinant of Q is either 1 or -1. The orthogonal matrices with determinant 1 will form a subgroup known as the special orthogonal group denoted by SO(n) which consists of all the direct isometrics of $O(n)$ and preserve the orientation of the space. SO(n) is a normal subgroup of $O(n)$. SO(2) is the special orthogonal group for $n = 2$. It is an orientation-preserving subgroup and is isomorphic to the circle group. The isomorphism produces the complex number $\exp(i\phi) = \cos(\phi) + i \sin(\phi)$ of absolute value 1 to the special orthogonal matrix $\begin{bmatrix} \cos\phi & -\sin\phi \\ \sin\phi & \cos\phi \end{bmatrix}$.

2.2.2 *Reflections*

An useful application of the orthogonal group is reflections. Reflections can be represented as elements of $O(n)$ group whose canonical form is $\begin{bmatrix} -1 & 0 \\ 0 & I \end{bmatrix}$ where $I = (n-1) \times (n-1)$ identity matrix and the zeros are columns of row zero matrices. This means that reflection is a transformation that transform the space in its mirror image with respect to a hyperplane. This is for one dimension. For two dimensions, each rotation is considered as the product of two reflections or a rotation of angle θ is the product of two reflections whose axes have an angle $\theta/2$. For every element of $O(n)$ is the product of a maximum of n reflections.

2.2.3 *Symmetry Group of Spheres*

The orthogonal group $O(n)$ is the symmetry group of all objects with spherical symmetry if the origin is chosen at the centre. It is the symmetry group of the $(n-1)$ sphere. For $n = 3$, or $O(n2)$ is the symmetry group of a sphere.

2.3 Unitary Group

The unitary group denoted $U(n)$ is a group of $n \times n$ unitary matrices with matrix multiplication as the group operation. It is also a subgroup of the general linear group GL(n, c). When $n = 1$ or $U(1)$, this corresponds to the circle group consisting of all complex numbers with absolute value 1 under multiplication. $U(n)$ is a real Lie group of dimension n^2 with the Lie algebra of $U(n)$ consists of $n \times n$ skew-Hermitian matrices with the Lie bracket given by the commutator.

The determinant of a unitary matrix is a complex number with norm 1. This determinant gives a group homomorphism

$$\text{Det} : U(n) \to U(1) \tag{2.5}$$

The set of unitary matrices with determinant 1 is the kernel of this homomorphism. This subgroup is known as the special unitary group or SU(n). This provides a short exact sequence of Lie group:

$$1 \to \text{SU}(n) \to U(n) \to U(1) \to 1 \tag{2.6}$$

From (2.6), one can view $U(1)$ as the subgroup of $U(n)$ that are diagonal with $e^{i\theta}$ in the upper left corner and 1 on the rest of the diagonal. Hence, $U(n)$ is a semi-direct product of $U(1)$ with SU(n).

The SU(n) groups have wide applications in physics such as in the standard model of particle physics with SU(2) in the electroweak interaction and SU(3) in quantum chromodynamics. With $n = 1$, the SU(1) is the trivial group with only a single element SU(2) group consists of 2×2 unitary matrices with determinant 1. It can be in this form:

$$U = \begin{pmatrix} a & b \\ c & d \end{pmatrix} \tag{2.7}$$

with the conditions: $\det(U) = 1$ and $UU^\dagger = 1$, one reduces tthe matrix U to

$$U = \begin{pmatrix} a & b \\ -b^* & a^* \end{pmatrix} \tag{2.8}$$

where the asterisk stands for complex conjugate. This is a two-dimensional representation of SU(2) group. The meaning of unitary matrix is as follows: a complex square matrix U is unitary if its conjugate transpose U^* is also its inverse. That is

$$U^*U = UU^* = I \tag{2.9}$$

In physics, the Hermitian adjoint of a matrix is denoted by a dagger (†) and the Eq. (2.9) becomes

$$U^{\dagger}U = UU^{\dagger} = I \tag{2.10}$$

The identity matrix I sometimes called a unit matrix of size n is the $n \times n$ square matrix with ones on the main diagonal and zeros elsewhere. It is denoted by I_n. Examples are:

$$I_1 = [1],\quad I_2 = \begin{bmatrix} 1 & 0 \\ 0 & 1 \end{bmatrix},\quad I_3 = \begin{bmatrix} 1 & 0 & 0 \\ 0 & 1 & 0 \\ 0 & 0 & 1 \end{bmatrix} \text{ etc.}$$

2.4 Lie Algebras [1]

Lie algebras were introduced by Marius Sophus Lie, the Norwegian mathematician in the 1870s to study infinitesimal transformations. It was also independently discovered by Wilhelm Killing. The name of Lie algebra was given by Herman Weyl in the 1930s. Lie algebras are closely related to Lie groups. Any Lie group gives rise to a Lie algebra. This is its tangent space at the identity. Lie groups are also smooth manifolds. For any finite-dimensional Lie algebra over real or complex numbers, there is a corresponding connected Lie group which is unique up to finite coverings. From this correspondence, one is able to study the classification and structure of Lie groups in terms of Lie algebras. The representations of Lie algebras are used extensively in physics.

A simple example is as follows. Take the space of three dimensional vectors $g = R^3$. The bracket operation is defined by the cross-product $[x, y] = x \times y$. Here $x \times y = -y \times x$ and so is skew-symmetric. It also satisfies the Jacobi identity:

$$x \times (y \times z) = (x \times y) \times z + y \times (x \times z) \tag{2.11}$$

This produces the Lie algebra of the Lie group of rotations of space. Each vector $v \in R^3$ represents an infinitesimal rotation. The Lie bracket is a measure of the non-commutativity between two rotations.

2.4.1 *Definition of Lie Algebra*

A Lie algebra is defined as a vector space g over some field F together with a binary operation [...,...]: $g \times g \rightarrow g$ called the Lie bracket satisfying the following axioms:

The Jacobi Identity

$$[x, (y, z)] + [z, (x, y)] + [y, (z, x)] = 0 \tag{2.12}$$

for all *x*, *y*, *z* in *g*.

Bilinearity

$$[ax + by, z] = a[x, z] + b[y, z] \tag{2.13}$$

$$[z, ax + by] = a\{z, x] + b[z, y] \tag{2.14}$$

for all elements *x*, *y*, *z* in *g* and all scalars *a*, *b* in *F*.

Alternatively,

$$[x, y] = 0 \tag{2.15}$$

for all *x* in *g*.

With the use of bilinearity to expand the Lie bracket, $[x + y, x + y]$ and with alternativity, it shows that $[x, y] + [y, x] = 0$ for all elements *x*, *y* in *g*. This shows that alternativity and bilinearity together imply.

Anticommutativity

$$[x, y] = -[y, x] \tag{2.16}$$

for all elements *x*, *y* in *g*.

Anticommutativity implies alternativity if the field's characteristic is not 2.

2.4.2 Relation to Lie Groups

Lie algebra is used as a means to study Lie groups. Lie algebra is related to Lie group can be shown by the fact that any Lie group gives rise to a canonically determined Lie algebra (the tangent space at the identity). Conversely, there exists a corresponding connected Lie group *G* with Lie algebra *g* for any finite-dimensional Lie algebra *g*. Any two Lie groups with the same Lie algebra are locally isomorphic and have the same universal cover although the Lie group is not determined uniquely. An example is that the special unitary group SU(2) and the special orthogonal group SO(3) give rise to the same Lie algebra which is isomorphic to $R^{3.}$ with the cross-product but SU(2) is a simply connected twofold cover of SO(3).

The correspondence between Lie groups and Lie algebras can be used in several ways such as in the classification of Lie groups and the related matter of the representation theory of Lie groups. Every representation of any Lie group induces a representation of the group's Lie algebra and conversely every representation of a Lie algebra lifts uniquely to a representation of the corresponding simply connected Lie group. These representations are in one-to-one correspondence. Hence, one can settle the problem of representations of the group by knowing the representation of a Lie algebra.

2.4.3 Abelian Group and Non-abelian Group [2]

An abelian group is a group in which the group operation is commutative. That is the result of applying the group operation to two group elements does depend on the order in which they are written. These are the groups that obey the axiom of commutativity. Abelian groups are named after Niels Henrik Abel. They generalize the arithmetic of addition of integers. A group G is said to be abelian if for every a, b $\in G$, $a \cdot b = b \cdot a$. That is, any two elements commute with each other. For the unitary group $U(n)$, for $n = 1$ or $U(1)$ is an example of an abelian group. $U(1)$ corresponds to the circle group, consisting of all complex numbers with absolute value 1 under multiplication. For $n > 1$, $U(n)$ is not abelian.

To fulfil the requirements of an abelian, the set and operation (A, .) must satisfy the following five axioms:

Commutativity

For all a, b in A, $a \cdot b = b \cdot a$.

Associativity

For all a, b, and c in A, $(a \cdot b) \cdot c = a \cdot (b \cdot c)$.

Inverse element

For each a in A, there exists an element b in A such that $a \cdot b = b \cdot a = e$, where e is the identity element.

Identity element

There exists an element e in A, such that for all element a in A, $e \cdot a = a \cdot e = a$.

Closure

For all a, b in A, the result of the operation $a \cdot b$ is also in A.

References

1. Sattinger, D.H., Weaver, O.L.: Lie Groups and Algebras with Applications to Physics, Geometry, and Mechanics. Springer-Verlag (1986)
2. Fuchs, L.: Infinite Abelian Groups, Pure and Applied Mathematics, vol. 36-II. Academic Press (1973)

Chapter 3
Linear Time-Reversal Acoustics

3.1 Principles: Spatial Reciprocity and Time-Reversal Invariance

Time-reversal acoustics is the branch of study which arises from the time-reversal symmetry property of the acoustic field. If one considers the Helmholtz wave equation in a lossless fluid medium:

$$\frac{\nabla^2}{\rho}p = \frac{1}{\rho c^2}\frac{\partial^2}{\partial t^2}p = 0 \tag{3.1}$$

where p = acoustic pressure, ρ = density, c = sound velocity = $\left(\frac{\kappa}{\rho}\right)^{1/2}$, and κ = compressibility.

Then time-reversal invariance means that if $p(\overrightarrow{r}, t)$ is an acoustics field solution of Eq. (3.1), then $p(-\overrightarrow{r}, t)$ is also a solution of this equation. Time-reversal invariance is valid only for a loss free medium. Also, it has to be noted that time-reversal invariance is not obeyed if the above equation contains odd order time derivative operator which happens for a frequency-dependent attenuation medium [1]. Unlike propagating waves, evanescent waves do not obey time-reversal invariance [2]. Evanescent waves have imaginary wave number compared to propagating waves which have real wave numbers. Evanescent waves occur when the incident sound wave is scattered by a medium with compressibility $\kappa(\overrightarrow{r})$ containing spatial frequency components of linear dimension smaller than the wavelength without a well-defined direction of propagation.

W. S. Gan, *Time Reversal Acoustics*,
https://doi.org/10.1007/978-981-16-3235-8_3

3.1.1 Principle of Time-Reversal Closed Cavity

The concept of the time-reversal cavity arises because it is not practical to measure the acoustic pressure $p(\overrightarrow{r}, t)$ in the whole three-dimensional volume during the time interval T, and then time reversed the transmission, that is $p(\overrightarrow{r}, T - t)$ in the whole volume where T is the time delay. This will amount to sampling the whole volume with transmit receive probes which is unrealistic. Using Huygens principle, one can reduce the operation from the 3D volume to a 2D surface because it states that one can predict the wavefield in any point of a volume from the knowledge of the field and its normal derivative. One can predict the acoustic pressure field inside the closed surface with this knowledge [3]. Using this as a starting point, one can consider the focusing on a target in an inhomogeneous medium as follows: One considers a point source in an inhomogeneous medium which emits a spherical wavefront which will be distorted after propagating in the medium. The object source and the inhomogeneities are surrounded by the closed surface. It is assumed that the acoustic pressure field and its normal derivative at any point on the surface can be measured. The next step, it is assumed that one can create secondary source such as monopole and dipole sources on the surface. These secondary sources correspond to the time-reversal of the signals measured during step one. Due to these secondary sources, a time-reversed signal backpropagates inside the surface and interacts with the inhomogeneities with a distortion of the wavefront. Under these situations, the time-reversed acoustic field will be focused on the initial source position [4, 5].

The analysis of the theory of the concept of closed time-reversal cavity can enable the understanding of the theoretical limitation of the principle of time-reversal. First one considers the first step of a self-focusing process with a closed time-reversal cavity of the object/source is located at the spatial coordinates origin then one has the following inhomogeneous Helmholtz wave equation:

$$\left(\nabla^2 - \frac{1}{c^2}\frac{\partial^2}{\partial t^2}\right)p(\overrightarrow{r}, t) = -\phi(t)\delta(\overrightarrow{r}) - A(\overrightarrow{r})p(\overrightarrow{r}, t) \tag{3.2}$$

where p = acoustic pressure, δ = Dirac delta function representing distribution in the 3D space and $\phi(t)$ = function of time, describing temporal variation of the source excitation. The second term on the right hand side of equation takes account of the inhomogeneities of the medium related to the compressibility and density of the medium [6]. $A(\overrightarrow{r})$ is a linear operator on the acoustic pressure $p(\overrightarrow{r})$. This second term can be considered as a secondary source term because it is the source of scattered sound generated by the interaction of the propagating sound field $p(\overrightarrow{r}, t)$ and the inhomogeneities. The excitation function $\phi(t)$ is zero for $t < 0$ and for $t > T$ where T is the time delay. This is based on causality and the temporal source excitation only exists on a finite time interval.

For a free unbounded medium, the solution of Eq. (3.2) can be given by [6]:

$$P(\vec{r},t) = \int\limits_V \left[\delta(\vec{r_0})\phi(t) + A(\vec{r_0})p(\vec{r_0},t)\underset{t}{*}G_d(\vec{r},\vec{r_0},t)d^3\vec{r_0}\right] \quad (3.3)$$

where V = volume enclosing both primary and secondary sources, $\underset{t}{*}$ = time convolution operator, $G_d(\vec{r},\vec{r_0},t)$ = free space Green's function corresponding to the diverging spherical wave. It satisfies the following propagating equations:

$$\left(\nabla_0^2 - \frac{1}{c^2}\frac{\partial}{\partial t^2}\right)G_d(\vec{r},\vec{r_0},t) = -\delta(t)\delta(\vec{r} - \vec{r_0}) \quad (3.4)$$

Equation (3.4) has the solution:

$$G_d(\vec{r},\vec{r_0},t) = [1/(4\pi/\vec{r} - \vec{r_0}/)]\delta\big(t - (/\vec{r} - \vec{r_0}/)/c\big) \quad (3.5)$$

The converging spherical wave is represented by

$$G_c(\vec{r},\vec{r_0},t) = [1/(4\pi/\vec{r} - \vec{r_0}/)]\delta\big(t + (/\vec{r} - \vec{r_0}/)/c\big) \quad (3.6)$$

and the diverging spherical wave is represented by the propagation equation:

$$\left(\nabla_0^2 - \frac{1}{c^2}\frac{\partial}{\partial t^2}\right)G_c(\vec{r},\vec{r_0},t) = -\delta(t)\delta(\vec{r} - \vec{r_0}) \quad (3.7)$$

Substitution of (3.5) into (3.3), then

$$p(\vec{r},t) = \frac{1}{4\pi/\vec{r}/}\phi\left(t - \frac{/\vec{r}/}{c}\right) + \int\limits_V \left[A(\vec{r_0})p(\vec{r_0},t)\right]\underset{t}{*}G_d(\vec{r},\vec{r_0},t)d^3\vec{r_0} \quad (3.8)$$

Here, the closed cavity is represented by the recording surface S which surrounds the object source and the inhomogeneities. The cavity does not interfere with the propagating sound field. The acoustic pressure field $p(\vec{r},t)$ and its normal derivative at any point of the surface on the cavity can be measured for a finite time interval [0, T].

The closed time-reversal cavity cannot be experimentally realized. It can be considered as an extension of the time-reversal mirror theoretically. It is not possible to generate only the time interval of an acoustic field as this time reversal is always followed by a backpropagation of the initial field. This causes a limitation on the self-focusing process.

To study the self-focusing in an inhomogeneous medium, one uses the propagation equation taking account of the spatial variation of the density and compressibility of the medium. One assumes that the inhomogeneities to be located in a specific region of space R in a finite volume near the origin. This equation is given by [6]:

$$\left(\nabla^2 - \frac{1}{c^2}\frac{\partial^2}{\partial t^2}\right)p(\vec{r},t) = \frac{1}{c^2}\Upsilon_\kappa(\vec{r})\frac{\partial^2}{\partial t^2}p(\vec{r},t) + \vec{\nabla}\cdot\left[\Upsilon_p(\vec{r})\vec{\nabla}p(\vec{r},t)\right] \tag{3.9}$$

where c = sound velocity $= \left(\frac{\kappa}{\rho}\right)^{1/2}$, and Υ is defined as $\Upsilon_\kappa(\vec{r}) = (\kappa_i(\vec{r} - \kappa)/\kappa)$ and $\Upsilon_\rho(\vec{r}) = \left[\rho_s(\vec{r}) - \rho\right]/\rho_i(\vec{r})$.

3.2 Limitation of Time-Reversal Acoustics—The Time-Reversal Mirror

The closed time-reversal cavity is an ideal proposal for focusing through an inhomogeneous medium. However, it is only a theoretical concept and difficult to realize in experiment. The greatest difficulty lies in the surrounding of the focal region by a set of transducers. This gives rise to the concept of time-reversal mirror which is based on the probe of the region locating only on one side of the region of intent. This is the same as that of the pulse echo mode used in non-destructive testing and in medical ultrasound imaging where the focusing is from an array of transducers. To use the time-reversed mirror in time-reversing focusing through an inhomogeneous media, the following steps have to be followed. First is the transmission of an acoustic field from the source, a transducer array to the target through the inhomogeneous medium. The sound field will be scattered and distorted. The scattered sound pressure field is recorded by the transducer array and this is the second step. The final step is the synthesis of the time-reversed field by the transducer array. This acoustic pressure field after propagating through the inhomogeneous medium focuses onto the target. For weak inhomogeneous field, only single scattering is considered and first Born approximation can be applied. Then the time-reversed mirror is able to compensate completely the wavefront distortion and this has been experimentally verified [7].

For the situation of sound propagation in strong inhomogeneities, the limitation of time-reversal mirror compared to closed cavity will be shown. This is due to the presence of multiple scatterings and the time-reversal mirror is unable to cope with the measurement over the long duration necessary to take account of the multiple scatterings that decay slowly.

Besides this, the time-reversal mirror is unable to measure the radiation of the scattered field in all directions. Hence, it cannot measure the full information required to optimize the real time reversal. On the other hand, for a weak inhomogeneous medium where single scattering is needed and the recording of the time-reversal is necessary only in one plane of the time-reversal mirror, it is sufficient to obtain optimum focusing on the target.

3.3 Time Reversal of the Source—The Acoustic Sink—Defeating the Diffraction Limit

In order for time-reversal acoustics to beat the diffraction limit, one needs to replace the initial source by an acoustic sink. To have a complete time reversal of the sound field, it is necessary to time reverse also the source. Then one can beat the diffraction limit. For an exact time-reversal experiment, during the second step, the initial source must be replaced by an acoustic sink. The acoustic sink is the time-reversal transformation of the acoustic source. To understand the above proposal, one needs the concept of evanescent waves. First, a point like source with size smaller than the wavelength of the transmitted sound wave radiates a sound field which is a superposition of the propagating sound wave in all directions and a non-propagating evanescent sound wave decaying exponentially with distance to the source. If the closed cavity is located at the far field of the source, then the time-reversed sound field reemitted from the surface of the cavity will not contain the evanescent waves. With the correct timing, the acoustic sink can radiate exactly the evanescent waves that have been lost during the first step. Then the retransmitted time-reversed sound field will contain the evanescent waves necessary for the focusing to beat the diffraction limit. This has been experimentally demonstrated using an acoustic sink placed at the focal point and producing focal spots of size $\lambda/4$ [8].

3.4 Experimental Demonstration of Time-Reversal Acoustics

The time-reversal mirror will be used to demonstrate the time-reversal acoustics [7]. There are 64 programmable transmitters which work simultaneously and are connected to a 64-element transducer array. The time-reversed experiments are performed in a water tank. A time-reversed mirror consists of a one or two-dimensional transducer array. Each transducer element is connected to its own electronic circuitry consisting of A/D converter, receiving amplifier, storage memory, and the most important component, the programmable transmitter able to synthesize time-reversed version of the signal stored in the memory. The time-reversal experiment for focusing on a point target through an inhomogeneous medium consists of three steps. During the first step, the transducer array illustrates an angular sector that contains the target. During the second step, the reflections from the target which are distorted by the inhomogeneities are recorded by the transducer array. In the final step, which is the focusing step, the transducer array reemits the time-reversed field. When there are several targets instead of one target, then the time-reversal process can be iterated. This iterative mode enables the selective focusing on the most reflective target so that the final transmit beam can converge onto a small portion of this target.

There are three types of transducer arrays, the TRM1, TRM2, and TRM3 used in the time-reversal experiment. The time-reversal mirror is a one-dimensional prefocused cylindrical time-reversal mirror made of a 64 element linear array. The TRM2 then is a one-dimensional time-reversal mirror comprising a linear array of 128 transducers suitable for medical imaging applications. The TRM3 is a part of a 2D curved transducer array. There are 64 transducers. For the focusing experiments, active point sources or passive point targets are used. The needle point hydrophones are used as active transmitting point sources. Both the point source and the cylindrical transducer array are immersed in a water tank. The acoustic pressure transmitted by the hydrophone was recorded by the 64 elements of the TRM1. The time-reversed experiment consisted of two stages. Stage one the individual signal recorded by each transducer was time reversed and transmitted individually. The hydrophone was used as a receiver and to monitor the pressure field at the initial location of the source. In the second stage, all the transducer elements simultaneously retransmit the time-reversed field, with a constructive interference of all the individual waves. The TRM focusing is self-adaptive and the geometrical distortion of the array was automatically compensated.

To perform the time-reversal mirror in the iterative mode, the TRM2, a 1D plane mirror will be used. Two different wires, copper wire and brass wire, were located below the transducer array. They were parallel to the long axis of each array element and located on both sides of the array axis. The experiment was carried out in two stages. During the first stage, a single transducer element located at the centre of the transducer array transmitted an illuminating beam with an angular sector containing the two wires. Although the single element has a wide directivity pattern, each single element emits a weak pressure level. The transmitted peak pressure can be increased with the programmable transmitters. The transducer element was driven by the electrical input signal matched to the acoustoelectric response. The reflection from the two targets was recorded after the first illumination. These recorded signals were time reversed and retransmitted. The measurements showed that the acoustic pressure field achieves a higher value at the positions of the copper wire which has a larger scattering cross-section than the brass wire. This process was iterated with the new reflections from the wire recorded, time reversed and retransmitted. After the iteration, the brass wire which reflects less energy received a weaker time-reversed wave. The brass wire is no longer illuminated after the last illumination. The experiments demonstrate the capability of the time-reversed mirror (TRM) iterative mode to select the most reflective target in a multitarget medium. This is a learning process that selects the one coming from the most important reflector. The process can be complicated by target resonance. This shows that the time-reversal mirror can be used to focus, through inhomogeneous media on a reflective target that behaves as an acoustic source after being insonified. Focusing is done on point targets through different aberrating media. The time-reversed focusing technique can compensate for all the distortions whatever the TRM-aberrator distance. For a medium containing several targets, the time-reversed process can be iterated to focus on the most reflective one.

References

1. Nachman, A., Smith, J., Wang, R.: An equation for acoustic propagation in inhomogeneous media with relaxation in losses. J. Acoust. Soc. Am. **88**, 1584–1595 (1990)
2. Nieto-Vesperinas, M., Wolf, E.: Phase conjugation and symmetries with wave fields in free space containing evanescent components. J. Opt. Soc. Am. **2**(9), 1429–1434 (1985)
3. Porter, R.P., Devaney, A.J.: Generalized holography and the inverse source problem. J. Opt. Soc. Am. **72**, 327–330 (1982)
4. Cassereau, D., Fu, W., Mink, F.: Limits of self-focusing using closed time-reversed cavities and mirrors-theory and experiment. In: Proceedings of IEEE Ultrasonics Symposium, Hawaii, Dec 1990, pp. 1613–1618
5. Cassereau, D., Fink, M.: Time reversal of ultrasonic fields. Part III. Theory of the closed time-reversal cavity. IEEE Trans. Ultrason. Ferroelectr. Freq. Control **39**(5), 579–592 (1992)
6. Morse, P.M., Ingard, K.U.: Theoretical Acoustics. McGraw Hill, New York (1968)
7. Wu, F., Thomas, J.L., Fink, M.: Time reversal of ultrasonic fields. Part II: experimental results. IEEE Trans. Ultrason. Ferroelectr. Freq. Control **39**(5), 567–578 (1992)
8. De Rosny, J., Fink, M.: Overcoming the diffraction limit in wave physics using a time-reversal mirror and a novel acoustic sink. Phys. Rev. Lett. **89**, 12401 (2002)

Chapter 4
Nonlinear Time-Reversal Acoustics

4.1 Principle of Nonlinear Time-Reversal Acoustics

Nonlinear time-reversal acoustics is an extension of linear time-reversal acoustics which was pioneered by Fink and his group [1]. There are two approaches to nonlinear time-reversal acoustics. One is to start with using finite amplitude sound wave as the working wave. The other approach is to use a linear sound wave and incident on objects with nonlinear elastic properties and providing nonlinear effects. Here, one will start with the first approach. Works on the use of finite amplitude wave as working wave in time-reversal acoustics have been done by Hedberg [2], Cunningham et al. [3], and Tanter et al. [4]. The simplest nonlinear acoustics wave equation representing finite amplitude wave is the Burgers' equation. The solutions of Burgers' equation are very complex. So, we have to start by examining the time-reversal symmetry property of the nonlinear finite amplitude sound wave; Hedberg [2] is of opinion that the finite amplitude sound wave obeys time-reversal invariance only when there is no dissipation and when there is no production of shock wave. There must be dissipation free propagation with no shock wave formed. The shock waves cause the loss of information in the finite amplitude sound wave. Time invariant property of the sound wave means that after receiving the sound wave, it is time reversed and retransmitted and the same original wave is received at the source.

Tanter et al. [4] also studied time reversal invariance of finite amplitude sound wave propagation experimentally. The shock wave aspect is in agreement with Hedberg [2]'s paper. They found that when the propagation distance is before shock wave formation the time-reversed invariance is obeyed. Here, the distortion of the sinusoid by nonlinearities during forward propagation will progressively recover its initial shape after the backward propagation. When the propagation distance is beyond shock wave formation, then time reversal invariance is broken.

Time reversal invariance means that at each position the time waveforms during forward and backward propagation are the time reversal of one another. Tanter et al. [4] analysed nonlinear sound wave propagation with the simplest case of the propagation of 1D nonlinear plane wave in an ideal fluid. The Burgers' equation can be

W. S. Gan, *Time Reversal Acoustics*,
https://doi.org/10.1007/978-981-16-3235-8_4

represented by the following set of equations for mass and momentum conservation, respectively:

$$\frac{\partial \rho}{\partial t} + \rho\frac{\partial v}{\partial x} + v\frac{\partial \rho}{\partial x} = 0 \tag{4.1}$$

$$p\left(\frac{\partial v}{\partial t} + v\frac{\partial v}{\partial x}\right) + \frac{\partial \rho}{\partial x} = 0 \tag{4.2}$$

where v = sound speed, ρ = density of medium, and since ideal fluid, entropy remains constant throughout. As the propagation is nonlinear, there will be interaction between two waves propagating towards $x > 0$ and towards $x < 0$ and the local sound speed will be modified. By substituting in condition for finding exact solution of Euler's equation and by introducing $\beta = \frac{B}{2A}$ where B/A = nonlinearity parameter, equations (4.1) and (4.2) are converted into:

$$\frac{\partial v}{\partial t} + (c_0 + \beta v)\frac{\partial v}{\partial x} = 0 \tag{4.3}$$

$$-\frac{\partial v}{\partial x} + (c_0 - \beta v)\frac{\partial v}{\partial x} = 0 \tag{4.4}$$

where c_0 = sound speed at equilibrium.

Equations (4.3) and (4.4) can be converted into the inviscid Burgers' equation for sound propagating into coordinates system moving with c_0. Equation (4.3) describes forward propagation with the boundary condition of $x = x_0$ before reaching the shock distance. Equation (4.4) describes the backward or reversed propagation towards negative directions. The boundary condition at $x = x_0$ is given by

$$g(t) - v(T - t) \tag{4.5}$$

where T = delay to ensure causality and $g = \frac{R-v}{2}$ with

$$R(\rho) = \int_{\rho_0}^{\rho} c(\rho)\frac{\mathrm{d}\rho}{\rho} \tag{4.6}$$

and (4.6) is known as Riemann invariant introduced by Riemann [5] to obtain an exact solution of Euler's equation. The time-reversed symmetry is obeyed in the nonlinear region before $x = x_0$ as $v(x, t)$ is solution of Eq. (4.3) and $-v(x, T - t)$ is the solution of Eq. (4.5).

For propagating distance beyond $x = x_0$, there will be shock waves formation and dissipation and time-reversal symmetry is broken. This happens even in an ideal fluid.

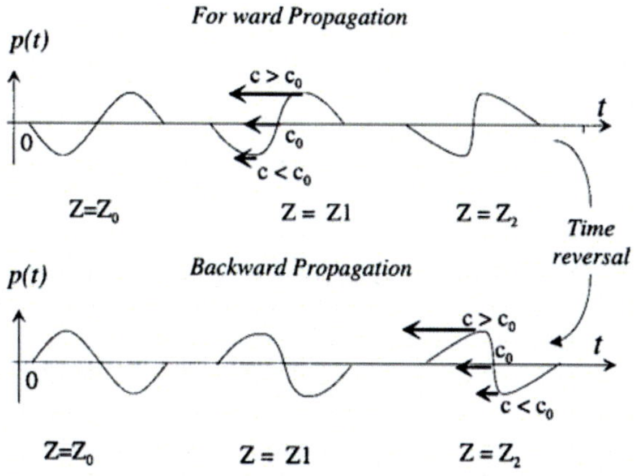

Fig. 4.1 A time-reversal experiment for a one-dimensional nonlinear propagation (after Tanter et al. [4])

Experimentally to check the validity of the time-reversal symmetry, Tanter et al. [4] initiated an experiment (Fig. 4.1) to test for cases of propagating distance within the boundary condition of $x = x_0$ and beyond $x = x_0$. For simplicity, the 1D plane wave is used. As a start, they synthesize a plane wave propagating along the mirror aperture. The synthesis is done by scanning the plane of time reversal using a point source at various positions. The experiment is performed in the nearfield using a plane piston consisting of 61 transducers with working central frequency of 1 MHz. The piston is immersed in water.

The transducers are arranged to form an hexagonal pattern. The ultrasound wave generated by each transducer is channelled through an electronic network with own programmable transmitter. The total signals from all the 61 transducers are computed. The next step is to time reverse all these 61 signals and use them to drive the transducers. A plane wavefront can then be produced in the control plane. A difficulty of producing time-reversal waves by using the time-reversal mirror is due to the requirement of devices with very broad bandwidths to reproduce the harmonics generated by signal distortion during forward propagation, but the ultrasonic transducer of the time-reversal mirrors has narrow bandwidths. This problem can be solved by using a pressure release interface for reflection purpose. This can be used for nonlinear sound wave propagation because it has the behaviour of the antisymmetry of a signal. However, the time-reversed signal is unable to be re-emitted with the time delay T. This means that both the incident and the reflected waves have to interact close to the interface. Besides this, the nonlinear interaction between forward and backward sound waves is very small compared with the distortions from the self-interaction of one-way waves. Tanter et al. [4] used a Mylar sheet stretching on a hollow cylindrical shell to produce the water–air interface.

Experimentally, for the purpose of a 1D plane, they checked the time reversibility before reaching $x = x_0$ at each point of propagation by showing that the time waveforms for forward and backward propagations should be the time reversal of one another. Here the shock distance of 1.1 m larger than the propagation distance, Mylar sheet is used. They found that time-reversal invariance is obeyed as forward and backward signals almost superpose and that time reversal enables the nonlinear distortion of the wave profile to be balanced. The backward propagation has also reversed the transfer of energy from the fundamental to the harmonics during forward propagation. Time-reversal symmetry is further confirmed that near the emitting array the backward propagating signal has perfectly returned to its original shape.

Beyond $x = x_0$, time reversing a compression shock will produce an expansion shock instead which will persist and cause instability and does not superpose with the forward propagating signal. Dissipation and nonlinear distortion will occur. Time-reversal invariance is broken due to shock wave formation and dissipation.

Tanter et al. [4] experimentally checked the time-reversal invariance for the shock wave propagation. Experimentally, they found that during backpropagation shocks are converted into expansion shocks that are unstable and initial waveforms cannot be recovered. This is known as irreversible shocks. This expansion shock also causes energy loss. During backpropagation, the forward shock waves are converted into backward expansion fans and stable compression shocks have been time reversed into unstable expansion shocks. The same experimental procedure is used in the checking of broken time-reversal invariance in propagation with shocks as that used in the confirmation of time-reversal invariance for 1D plane wave before reaching the shock boundary but with a shock distance of 0.3 m smaller than the propagation distance.

4.2 Second Definition of Nonlinear Time-Reversal Acoustics

Here instead of defining nonlinear time-reversal acoustics (TRA) as TRA using nonlinear finite amplitude sound wave, there is another school of researchers who interpret nonlinear TRA as the application of TRA to nonlinear scatterers. The purpose is the imaging of nonlinear scatterers. This is the application of the time-reversal mirror (TRM) to the imaging of nonlinear scatterers. This is applied to damage detection as damage produces nonlinear effects. Damage in the form of cracks and flaws induces nonlinear effects and produces harmonics, whereas voids and sidewalls do not induce nonlinear response. Focusing time-reversal signals onto nonlinear scatterers and analysing the nonlinear response will enable the imaging of the cracks and damages. There is nonlinear interaction of the time-reversed focused signal and the damage. The large localized amplitude of the sound wave generated by time-reversal focusing is useful for the investigation of nonlinear effects. Experimentally, it is carried out using two frequencies the sum and the difference frequencies. Waveform in the cracked region contains the nonlinear signature of sum and difference frequency spectral components. While the presence of nonlinear components of the waveform in the damaged area did not come as a surprise, we were, however, surprised at the degree of amplification of the entire focused signal on the nonlinear scatterer.

In the use of time-reversal acoustics for focusing and imaging of nonlinear scatterers, the technique of nonlinear elastic wave spectroscopy (NEWS) is commonly used. It is for the determination of the degree of presence and the degree of nonlinearity in elastic medium. The nonlinearity behaviour of materials is due to various reasons such as dislocation, harmonics, and nonlinear interaction between time-reversal signals and damage. The nonlinear effects are present locally or globally throughout the medium.

Nonlinearity produces intermodulation distortions and harmonics or additional frequency contents not present in the sound wave generated by the original source. The mechanism of the interaction of the time-reversal sound wave with the nonlinear scatterer is rather complex. The localized distortion of the sound wave gives rise to harmonics, intermodulation, and wave mixing [6]. There are two forms of the nonlinear scatterers: they can be localized cracks, delamination, thermal damage, etc., in an elastically linear medium or present globally throughout the medium such as in rocks, some form of metals, and ceramics [6, 7]. The detection of nonlinearity is an extremely sensitive diagnostic method for damage [7–9]. More complex and nonlinear characteristics of elastic wave propagation like third harmonics, etc., are starting to be also used for ultrasonic imaging, e.g. in medicine, providing more sensitive indicator of small body changes and irregularities. Nonlinear parameters obtained by NEWS NDT techniques can give us better information about incipient damage zones than linear ones. The main advantage of NEWS consists not only in higher sensitivity but also in the ability to detect defects smaller than the ultrasonic wavelength.

The time-reversal focusing will produce high-intensity sound wave useful for the imaging of nonlinear effects which include generation of harmonics, interaction of sound wave with nonlinear scatterers, etc. To study the theory of time-reversal signal scattering from nonlinear elastic scatterers, Dos Santos et al. [10] have derived the nonlinear parameter *B*/*A* from the measurements of signals generated by the combination of frequencies generated from the interaction of time-reversal sound waves with the nonlinear scatterers.

The Westervelt equation [11] is used which contains the nonlinear parameter $\epsilon = 1 + B/2A$. The Westervelt equation [11] is written as

$$\nabla p - \frac{1}{c_0^2}\frac{\partial^2}{\partial t^2}p = -(\epsilon/\rho_0 c_0^4)\frac{\partial^2}{\partial t^2}p^2 \tag{4.7}$$

There are the primary pressure p_{p} and the secondary pressure p_{s} generated by nonlinear scattering. Rewriting (4.7) in terms of p_{p} and p_{s}, one has

$$\nabla p_{\mathrm{s}} = -\frac{1}{c_0^2}\frac{\partial^2}{\partial t^2}p_{\mathrm{s}} = -(\epsilon/\rho_0 c_0^4)\frac{\partial^2}{\partial t^2}p^2 \tag{4.8}$$

where

$$\mathrm{d}p_{\mathrm{s}} = -\frac{\rho_0}{4\pi r} q(t - r/c_0) \text{ and } q = -(\epsilon/\rho_0^2 c_0^4) \frac{\partial^2}{\partial t^2} p_{\mathrm{p}}^2 \tag{4.9}$$

The volume element where interaction takes place is given by $\mathrm{d}V$, and q is the density of the spherical wave generated by this volume element. Difference frequency and sum frequency are produced by the interaction of the two waves of frequencies ω_1 and ω_2. Their source densities are given by

$$q_- \text{ for difference frequency} = \left[\epsilon(\omega_1 - \omega_2)^2/2\rho_0^2 c_0^4\right] p_1 p_2^* \tag{4.10}$$

$$q_+ \text{ for sum frequency} = \left[\epsilon(\omega_1 + \omega_2)^2/2\rho_0^2 c_0^4\right] p_1 p_2 \tag{4.11}$$

Then, the total radiated field can be given as

$$p_\pm == \left[\epsilon(\omega_1 + \omega_2)^2/2\rho_0^2 c_0^4\right] \iiint p_{1,2}[\exp(-\mathrm{i}k_{1,2}r)]/2\,\mathrm{d}V \tag{4.12}$$

and

$$k_{1,2} = (\omega_1 \pm \omega_2)/c_0 \tag{4.13}$$

The nonlinear time-reversal focused acoustic field is almost a spherical converging wave of the form

$$p_{1,2} = \sin(k_{1,2}r)/k_{1,2}r \tag{4.14}$$

Substituting (4.13) into (4.12) will produce

$$p_\pm = \alpha(\omega_1 + \omega_2)^2 \epsilon p_1(r = 0) p_2(r = 0) d^3/24\rho_0 c_0^4 r \tag{4.15}$$

where d = diameter of the focused area of interaction. Here, α is a parameter taking into account the nonuniform distribution of the primary field and is approximately equal to 1 when different frequency wavelengths are much less than d.

4.3 Experimental Verification of Nonlinear Time-Reversal Acoustics

Before the arrival of the application of time-reversal acoustics to study nonlinear effect from scattering of time-reversal signal by nonlinear elastic objects, in the 1980s, there are experimental works on the use of acoustic tomography to study nonlinear parameters of tissues such as the *B/A* nonlinear parameter [12]. However, these methods suffer from the difficulty to be used in actual clinical situations. In place of these methods is the use of time-reversal acoustics. Time-reversal signals can be focused on the part of tissue with nonlinear interaction, and highly concentrated acoustic energy can take place in heterogeneous media. The measurement of nonlinear parameter is given by the nonlinear scattering from the concentrated area. Nonlinear acoustical imaging can be realized by scanning the focused region in the specimen.

For the experimental verification of nonlinear time-reversal acoustics, the first step is to estimate the nonlinear response of an elastic material. This can be done by measuring the modulation of the incident sound wave ($f_{\rm hi}$) by vibration of low frequency $f_{\rm lo}$. Nonlinear wave mixing will arise in materials with localized damage such as a crack in a solid or distributed damage such as that in a rock. The spectral component will consist both sum and difference frequencies ($f_+ = f_{\rm lo} + f_{\rm hi}$ and $f_- = f_{\rm hi} - f_{\rm lo}$) and pure tones $f_{\rm lo}$ and $f_{\rm hi}$. On the other hand, an undamaged material will consist only pure tones. f_+ and f_- are additional frequencies. The sum and difference frequencies are considered as sidebands to the incident ultrasound frequency $f_{\rm hi}$.

The object used for the experiment of Ulrich and Johnson [13] is a doped glass block with a surface crack. There are two sources made of piezoelectric ceramics and were bonded to the surface of the glass block and opposite to the crack. A broadband laser vibrometer was used to measure the elastic response on the crack. A block diagram of the experimental setup is given in Fig. 4.2.

The two sources were driven by frequencies $f_1 = f_{\rm lo}$ and $f_2 = f_{\rm hi}$. The signals were measured directly for each source independently and then time reversed. Time reversed invariance will enable the resulting signal to be focused at the point of the original detector. The experimental procedure is as follows:

I. Source n generates signal $S_n(t)$ where

$$S_n(t) = U_n \cos(2\pi f_n t) \sin^2\left(\frac{\pi}{\tau} t\right) \tag{4.16}$$

where f_n = frequency generated from nth transducer with maximum amplitude U_n and duration τ.

II. The signal $r_{m,n}(t)$ is received at the focal point x_m of the laser vibrometer on the sample independently for each f_n.

III. The signal $r_{m,n}(t)$ is time reversed, i.e.

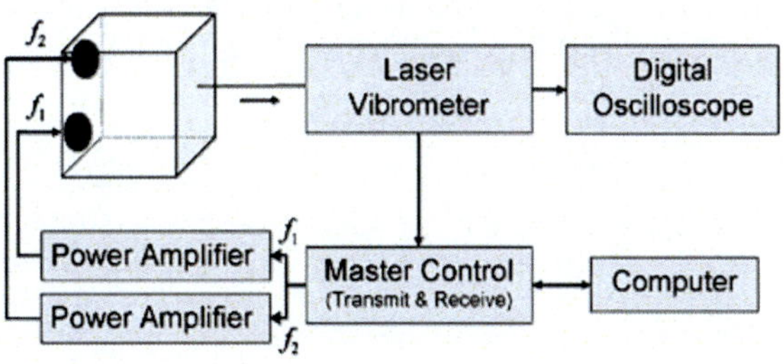

Fig. 4.2 Block diagram of the nonlinear TRA(NTRA) experimental system (after Ulrich and Johnson [13])

$$r_{m,n}(t) \rightarrow r_{m,n}(-t) \tag{4.17}$$

IV. The time-reversal signals $S_{m,n}(t)$ are amplified and simultaneously transmitted from their original source location,

$$S_{m,n}(t) = U_n r_{m,n}(-t)/\max(r_{m,n}(-t)) \tag{4.18}$$

V. The time-reversed focused signal $R_m(t)$ is recorded at the mth focal point of the laser vibrometer (x_m),

$$R_m(t) = \sum_{n=1}^{N} \int_0^t \mathrm{d}t' G(x_m t/x_n t') S_{m,n}(t') \tag{4.19}$$

where N = total number of source transducers, x_n = location of the nth transducer, $G(x_m t/x_n t')$ = Green's function taking the signal transmitted from location x_n at time t' to point x_m at a later time t. The integration is over the entire length of the waveform.

VI. The focal point of the laser vibrometer is then moved to the next desired location x_{m+1}. The whole process is then repeated for M total focal points.

In this experiment, the whole cracked area was scanned. The same f_1 and f_2 sources were used for each of the focal point. Graduated increments with consistent step size were used during scanning across the crack. The presence of nonlinear effects was studied by analysing the waveform of the focused signals saved at each focal point. First, this was done by comparing the complete set of unfiltered waveform as a function of position. Next, these waveforms were then filtered about the frequencies corresponding to the sum and difference frequencies and then compared as a function of positions. It was found that the amplitudes of the focused waveforms increased near the region of the crack and the sidelobe amplitudes also have same characteristics. Also, the nonlinear features exhibit narrower focusing. This indicates the extent of the crack boundaries. Besides this, the sum frequency and difference frequency behave differently inside the crack. With signal focused at different locations along the crack, the difference frequency waveforms are more structured in amplitude than the sum frequency waveforms. Also, they exhibit more sidelobes. The above techniques can be used to discriminate between linear and nonlinear scatterers and can ultimately lead to the imaging and characterization of damaged regions. The advantage of nonlinear acoustical imaging of nonlinear elastic scatterers is higher sensitivity and higher resolution compared with linear acoustical imaging.

References

1. Fink, M., Prada, C., Wu, F.: Self focusing in inhomogeneous media with time reversal acoustic mirrors. In: McAvoy, B.R. (ed.) Proceedings of IEEE Ultrasonics Symposium, vol. 2, pp. 681–686 (1989)
2. Hedberg, C.: Solving the inverse problems in nonlinear acoustics by backpropagation of the received signal. J. Acoust. Soc. Am. **101**(2), 3090 (1997) (Abstract)
3. Cunningham, K.B., Hamilton, M.F., Brysev, A.P., Krutyansky, L.M.: Time-reversed sound beams of finite amplitude. J. Acoust. Soc. Am. **109**(6), 2668–2674 (2001)
4. Tanter, M., Thomas, J.L., Coulouvrat, F., Fink, M.: Breaking of time reversal invariance in nonlinear acoustics. Phys. Rev. E **64**, 016602 (2001)
5. Hamilton, M.F., Blackstock, D.T.: Nonlinear Acoustics. Academic Press, New York (1998)
6. Guyer, R.A., Johnson, P.: Nonlinear mesoscopic elasticity: evidence for a new class of materials. Phys. Today **52**, 30 (1999)
7. Johnson, P.A., Sutin, A.M.: Slow dynamics and anomalous nonlinear fast dynamics in diverse solids. J. Acoust. Soc. Am. **117**, 124 (2005)
8. Van Den Abeele, K., Johnson, P.A., Sutin, A.: Non-linear elastic wave spectroscopy (NEWS) technique to discern material damage, Part I: nonlinear wave modulation spectroscopy. Res. Nondistr. Eval. **12**, 17 (2000)
9. Solodov, I., Wackerl, J., Pfleiderer, K., Busse, G.: Nonlinear self-modulation and subharmonic acoustic spectroscopy for damage detection and locations. Appl. Phys. Lett. **84**, 5386 (2004)
10. Dos Santos, S., Choi, B.K., Sytin, A., Sarvazyan, A.: Proceedings of the 8 eme Congres Francais d'Acoustique, pp. 359–362 (2006)
11. Westervelt, P.J.: Parametric acoustic array. J. Acoust. Soc. Am. **35**, 535–537 (1963)
12. Sato, T., Fukusima, A., Ichida, N., Ishikawa, H., Miwa, H., Igarashi, Y., Shimura, T., Murakami, K.: Nonlinear parameter tomography system using counter propagating probe and pump waves. Ultrason. Imaging **7**(1), 49–59 (1985)
13. Ulrich, T.J., Johnson, P.A.: Imaging nonlinear scatterers applying the time reversal mirror. J. Acoust. Soc. Am. **119**(3), 1514–1518 (2006)

Chapter 5
Non-reciprocal Acoustics

5.1 Meaning of Non-reciprocal Waves

In this chapter, we are investigating the situation when time-reversal acoustics fails. It has been shown that for fluid in motion, time-reversal symmetry is not upheld [1, 2]. For the last decade has seen the arrival of engineered acoustic devices with non-reciprocal behaviour such as the acoustic diode [3], acoustic isolator, and acoustic circulator [4]. These devices enable the sound wave to propagate only in one direction. Time-reversal acoustics arises from the time-reversal symmetry property of the acoustic fields. In a linear, inhomogeneous medium, there are two ways of breaking time-reversal symmetry. One is microscopic absorption loss which removes energy from the wave propagation. Another process depends on the properties of the medium described by a parameter or a set of parameters. An example of the parameter is the magnetic field.

Besides the linear medium, another way of breaking time-reversal symmetry is for large amplitude sound propagation in medium with dissipation and shock wave presence. In the subsequent section of this chapter, we will only describe the acoustic devices that will break the time-reversal symmetry and produce only unidirectional or non-reciprocal waves.

5.2 The Onsager-Casimir Principle [5] of Microscopic Reversibility

The Onsager reciprocal relations [6] is the root of time-reversal symmetry. Time-reversal symmetry is given rise by the symmetry property of the wavefield. It can be also considered from the symmetry property of the process which is a transport process and a transport phenomenon.

This process is based on the principle of microscopic reversibility. The Onsager reciprocal relations [6] is a transport theory. It can be derived from the theory of

W. S. Gan, *Time Reversal Acoustics*,
https://doi.org/10.1007/978-981-16-3235-8_5

fluctuation and the principle of microscopic reversibility. Start with two examples: tensor of heat conductance and Kelvin's relation between thermoelectric quantities, *N*. Bohr was first to demonstrate that a general class of reciprocal relations can be derived from this principle of microscopic reversibility without having to recourse to any particular model and he has applied his theory to numerous examples including those mentioned above, adding a new and very fundamental principle to fundamental thermodynamics. Onsager's reciprocal relations [6] is derived from *N* Bohr's microscopic reversibility which means invariant under transformation t to $-t$. It is derived from the assumption of microscopic reversibility and the fundamental laws of thermodynamics and the production of entropy. A general reciprocal relation applicable to transport processes such as conduction of heat and electricity and diffusion is derived from the assumption of microscopic reversibility.

5.3 Application of Onsager Reciprocal Relations

There are two examples illustrating the application of the Onsager reciprocal relations [6]. One is the heat conduction in crystals and the electricity conduction. Heat conduction in crystal is a heat transport process and a transport phenomenon. To describe the heat conduction in crystals, one has the following equation:

$$w_i = \sum_k L_{ik} \partial_k T \tag{5.1}$$

where $\partial_k =$ derivative with respect to x_k and x_1, x_2, x_3 the Cartesian coordinates x, y, z and $L_{ik} =$ conductance.

To apply Onsager's theory, one needs to include temperature fluctuations by dividing the solid into a number of cells and calculates the change of entropy in terms of the specific heat. To apply Onsager symmetrical relation to electrical conductance, one considers the situation of a conducting solid with four leads attaching at points A, B, C, and D with the condition that current I_1 entering at point A be equal to current leaving at point B and that current I_2 entering at point C be equal to current leaving at point D and V_1 and V_2 representing the potential differences between A and B and between C and D. This will be represented by the following equations:

$$V_1 = A_{11} I_1 + A_{12} I_2 \tag{5.2}$$

and

$$V_2 = A_{21} I_1 + A_{22} I_2$$

or

$$I_1 = a_{11} V_1 + a_{12} V_2 \tag{5.3}$$

and

$$I_2 = a_{21}V_1 + a_{22}V_2$$

From Onsager's reciprocal relations [6], one has

$$a_{12} = a_{21} \quad \text{and} \quad A_{12} = A_{21} \tag{5.4}$$

The above symmetry relation will hold so long as there is no magnetic field or

$$a_{12}(H) = a_{21}(-H) \tag{5.5}$$

where H = magnetic field.

Next one considers the Onsager reciprocal relations [6] for the electric resistance tensor:

$$E_i = \sum R_{ik}i_k \tag{5.6}$$

where E_i = tensor component of the electric field strength and i_k = current density components.

Then

$$R_{ik} = R_{ki} \tag{5.7}$$

if $a_{ik} = a_{ki}$.

It is due to the symmetry of all the four poles that enables the above symmetry relations for the electric resistance tensor.

5.4 Thermal Diode

The thermal diode is the predecessor of the acoustic diode. The inventor of the acoustic diode was inspired by this. It was invented by Li et al. [7]. It enables one direction heat flow. This is a better understanding of heat conduction and transport phenomenon. The thermal diode allows heat flow from one end to the other but prohibits the flow from the opposite direction. This is done by coupling two nonlinear lattices of particle chains using a harmonic spring with constant strength. The interface elastic constant is an important parameter as it controls right and left heat flow. The rectifying effect is achieved by varying the parameters which control the nonlinearity of the lattice. The Li et al. [7] paper is an improvement over the previous paper of Terraneo et al. [8].

To derive the rectifying mechanism, one considers the energy spectrum of the interface particles. The heat current is controlled by the match/mismatch of the energy spectra of the two interface particles. If the left end is in contact with the high temperature bath and the right end with the low temperature bath, then the phonon spectra of the two particles at the interface overlap in a large range of frequencies. In this case, the heat current will flow from the left end of the system to the right end of the system. On the other hand, if the left end is in contact with the lower temperature bath and the right end with the higher temperature bath, then the phonon spectrum of the right interface particle is mainly in the low frequency portion and that of the left interface particle is mainly in the high-frequency portion. In this case, there is almost no overlap in the phonon frequency and there will be no heat flow from right to the left. In order to maximize the rectifying effect, the overlapping of the phonon bands in the low temperature limit and for each segment of the system in the high temperature limit should be avoided. The temperature dependence of the phonon band is contributing to the rectifying effect. This is due to the nonlinearity of the potential energy.

Hence, it is possible to observe rectifying effect in any nonlinear lattice.

Since the rectifying effect depends on the coupling between two dissimilar lattices, the rectifying efficiency will depend on the number of nonlinear oscillations. By increasing the number of nonlinear oscillations and the coupling constant strength, the rectifying efficiency will be reduced.

5.5 Acoustic Diode

The concept of acoustic diode was inspired by the thermal diode [7]. Acoustic diode allows part of acoustic energy flux incident from one particular side to pass, but behaves like an insulator as the wave propagated in the opposite direction. Such a rectifying phenomenon occurs when frequency of incident wave fell within particular frequency bands. There is measured pressure dependence of transmitted acoustic energy flux with strong similarity to voltage–ampere relationship of electric diode. The rectifying effect is affected by ultrasound contrast agent concentration.

Sonic crystal as acoustic filter can yield different transmission property between fundamental and second harmonic waves. Nonlinear materials are introduced to destroy system symmetry. If acoustic waveforms come from proximal side of nonlinear medium (NLM) hit nonlinear material first, creating simple harmonic motion passes through the filter, then any sound from opposite direction at fundamental wave frequency is blocked before reaches the nonlinear medium (NLM). Therefore, asymmetric propagation of sound wave can be expected. This is known as acoustic rectification.

Nonlinear materials can be made of UCA suspension. UCA microbubble should have a relatively broad size distribution; it is possible to yield acoustic nonlinearity in a wide frequency range.

Acoustic diode can be applied to noise control. For the purpose of noise control, many recording studios, hotel rooms, and even some homes are built with materials to help reflecting or absorbing sound. However, mechanisms that truly control the direction of sound waves are still in their infancy. Researchers at the California Institute of Technology have created the first tunable acoustic diode, a device which allows in architectural acoustics. The system is based on a simple assembly of elastic spheres—granular crystals that transmit the sound vibration—that could be easily used in multiple settings, can be tuned easily, and can potentially be scaled to operate within a wide range of frequencies to enable its applications to reach beyond sound-proofing. It is necessary that the sharp transition between transmitting and non-transmitting states to be more effective at controlling the flow of sound waves. The team created a periodic system with a small defect that supports the quick change from an on to an off transmission state. Thus, the system is very sensitive to small variations of operational condition, like pressure and movement, enabling it useful in the development of ultrasensitive acoustic sensors for detecting sound waves. The system is able to operate at different sound frequencies and is capable of downshifting or reducing the frequency of the travelling signals as required.

The device consists of two parts: a layer of ultrasound gel and a lattice comprising alternating layers of water and glass. For sound to pass through the diode, sound waves must hit the gel layer first. The gel is a nonlinear medium that converts some of the energy of the acoustic wave into a wave of double the frequency. The lattice then screen out all frequencies except the second, higher frequency harmonics of the original wave, allowing some of the sound through, albeit at a higher pitch. However, if the original acoustic wave attempts to pass the other way back through the device, it will first encounter the lattice. Because this wave has not passed through the gel layer and thus has only the original frequency, it will be totally screened out by the lattice. Such a device would be useful for medical ultrasound application for cases which require extremely focused sound wave of specific frequencies.

The tunable acoustic diode of Li et al. [7] is sonic crystal based. The spatial inversion symmetry in the sonic crystal was broken. It is a completely linear system without any acoustic nonlinearity. The design of the acoustic diode consists of an asymmetric periodic corrugated sonic crystal shown in the lower panel of Fig. 5.1a which has a regular sonic crystal and a diffraction structure.

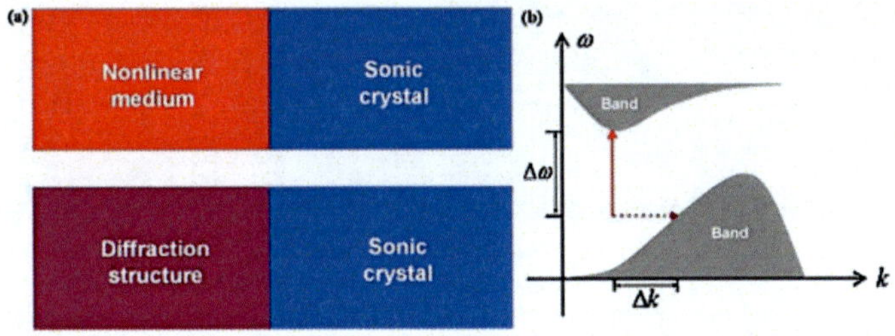

Fig. 5.1 **a** Illustration for two diode models, with the top one coupling nonlinear medium to SC and the bottom one using diffraction structure. **b** Schematic for transitions between different modes: the solid red line represents the transition with frequency change and the dashed purple line represents the transition between different spatial modes (after Li et al. [7]) (colour online)

Due to the diffraction structure, there is transition between two spatial modes with different spatial frequencies shown by the dashed purple line in Fig. 5.1b. The adjacent sonic crystal has intrinsic anisotropy of acoustic band structure. Hence, it acts as a spatial filter so that different spatial modes such as high-order diffraction modes of different parallel wave vectors will be prohibited or transferred in the sonic crystal. The physical structure of the acoustic diode given in Fig. 5.2a consisted of a two-dimensional sonic crystal formed of a square mesh and a corrugated diffraction structure. Both the diffraction structure and the sonic crystal were composed of square steel rods. Numerical simulation using finite element method was carried out to evaluate the unidirectional transmission property of the acoustic diode. The transmission spectra demonstrated unidirectional, nonreciprocal transmission efficiencies for the right incidence (RI) and the left incidence (LI) with a normally incident plane wave. The experiments were performed. The measurement results agreed with the

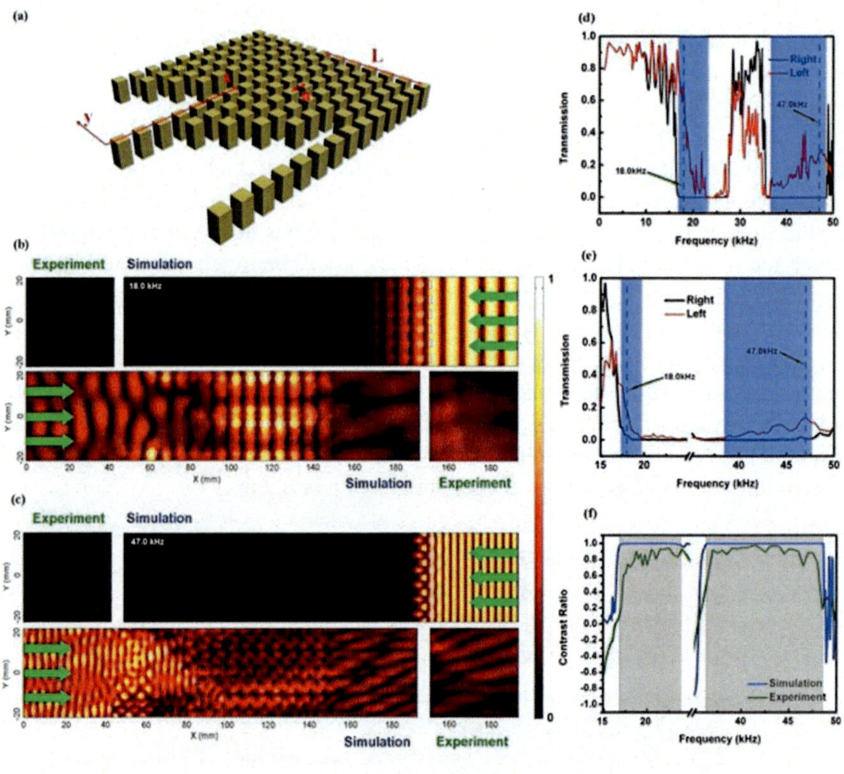

Fig. 5.2 **a** Schematic diagram of the SC-based acoustic diode which is periodic in y direction. **b, c** Simulated and experimental field distribution mappings with the incident wave frequency at 18.0 and 47.0 kHz, respectively. Green arrows indicate the propagation directions. **d, e** Numerically calculated and experimentally measured transmission spectra of LI and RI, respectively. Green arrows indicate the frequency at which field distribution is mapped. **f** Contrast transmission ratio of the acoustic diode (after Li et al. [7]) (colour online)

simulation results despite the limitation due to the finite size in the transverse direction and the slight imperfection of the plane wave source in the experiment. Due to the fact that the outgoing waves are not parallel to the incident waves, energy from normal incidence will be converted to other spatial modes that have different spatial frequencies to overcome the barrier from the directional band gap.

Li et al. [7] quantify the performance of the acoustic diode by defining the constant ratio R_c as

$$R_c = (T_L - T_R)/(T_L + T_R) \tag{5.8}$$

where T_L = transmission for left incidence and T_R = transmission for right incidence. The relative transmission weight between these two incident cases is given by the absolute value of R_c.

Besides the broken spatial symmetry given by the diffraction structure, the building block of the Li et al. [7] sonic crystal also includes the steel square rod as a broken rotational symmetry to the unit cell. This broken rotational symmetry can cause the change of the sonic crystal's band dispersion by mechanically rotating all the square rods with the same lattice configuration. This is due to different rotations significantly affect the effective scattering section for acoustic wave. Hence, it can effectively control the sound rectification by efficiently tuning the relative transmission weight between left incidence and right incidence with the acoustic diode by rotating the rods. The sound rectification effect can be effectively turned on or off in the acoustic diode by rotating the square rods of the sonic crystal. The tunable unidirectional transmission effect can also be influenced by different rotation angles. With a more systematic design of the square rod-based sonic crystal, the tunable acoustic diode can be constructed either in the first or second band or both. Hence, the frequency range of unidirectional transmission can be effectively controlled with different rotational angles of these square rods.

The following is the mechanism for the above unidirectional transmission phenomenon. In terms of Fourier series, and using a plane wave, the acoustic pressure field in the input half-space can be written as:

$$p(x, y) = \exp(\mathrm{j}k_0 x) + \sum_{n=-\infty}^{\infty} \rho_n \exp(\mathrm{j}\alpha_n y + \mathrm{j}\beta_n x) \tag{5.9}$$

The acoustic pressure in the output half-space can be written as

$$p(x) = \sum_{n=-\infty}^{\infty} \tau_n \exp(\mathrm{j}\alpha_n y + \mathrm{j}\beta_n x) \tag{5.10}$$

where α, β = wave numbers in the x- and y-direction, respectively, $\alpha_n = 2n\pi/L$ where n = diffraction order, $\beta_n = \sqrt{(k_0^2 - \alpha_n^2)^-}$.

ρ_n, τ_n = amplitudes of the nth order diffraction beams in the reflection and transmission mode, respectively.

These two diffraction terms are responsible for the mechanism of the acoustic diode. They represent the energy conversion among different diffraction orders. Different diffraction orders are produced by the difference of structure geometries on two sides. It is to be noted that only the diffraction orders locating within the propagating band of the sonic crystal can transmit the energy from one side to the other.

The acoustic diode of Li et al. [7] is a completely linear system. This has the advantages compared with other acoustic diodes using nonlinearities. These are much less power consumption, high conversion efficiency, and broadband operation.

5.6 Applications of the Acoustic Diode

The unidirectional effect of the acoustic diode can be used in architectural acoustics with applications to noise control such as to hotel rooms, recording studios and even some homes to be built with materials to help absorb or reflect sound. The acoustic diode which can control the direction of sound waves is applicable here which allows acoustic information to travel only in one direction at controllable frequencies. Acoustic diode is able to scavenge sound energy from undesired structural vibrations in machinery by controlling the flow of sound waves away from the machinery into a transducer which converts the sound wave into electricity. This technique can also shift the undesired frequencies to a range that enables a more efficient conversion into electricity. By enabling the unidirectional acoustic wave flow, the effect of interference from sound wave travelling in opposite direction can be eliminated. This can provide a higher image resolution during acoustical imaging. Acoustic diode enables the manipulation of vibrational energy for energy harvesting. Non-reciprocal materials have potential in the development of the acoustic equivalent of topological insulators with application in energy harvesting and vibrational energy isolation [9].

References

1. Morse, P.M., Ingard, K.U.: Theoretical Acoustics. McGraw-Hill, New York (1968)
2. Godin, O.A.: Reciprocity and energy theorems for waves in a compressible inhomogeneous moving fluid. Wave Motion **25**, 143–167 (1997)
3. Li, X.F., Ni, X., Feng, L., Lu, M.H., He, C., Chen, Y.F.: Tunable unidirectional sound propagation through a sonic-crystal-based acoustic diode. Phys. Rev. Lett. **106**, 084301-1–084301-4 (2011)
4. Fleury, R., Sounas, D., Haberman, M.R., Alù, A.: Nonreciprocal acoustics. Acoust. Today **11**(3), 14–21 (2015)
5. Casimir, H.B.G.: On Onsager's principle of microscopic reversibility. Rev. Mod. Phys. **17**, 343–350 (1945)

6. Onsager, L.: Reciprocal relations in irreversible processes. Phys. Rev. **37**(4), 405–426 (1931)
7. Li, B., Wang, L., Casati, G.: Thermal diode: rectification of heat flux. Phys. Rev. Lett. **93**, 184301 (2004)
8. Terraneo, M., Peyrard, M., Casati, G.: Phys. Rev. Lett. **88**, 094302 (2002)
9. Khanikaev, A.B., Fleury, R., Mousavi, H.S., Alù, A.: Topologically robust sound propagation in an angular momentum-biased graphene-like resonator lattice. Nat. Commun. **6**, 8260 (2015)

Chapter 6
Onsager Reciprocal Relations

6.1 Principle of Microscopic Reversibility-as the Root of Time-Reversal Acoustics

The principle of microscopic reversibility involves these aspects. Firstly, it states that the microscopic description of the dynamics of fields and particles in terms of the equation of motion of these particles have time invariance or obey time-reversal symmetry. Secondly, it provides a statistical description of the kinetics of macroscopic or mesoscopic system as an ensemble of elementary processes such as elementary transition, collision, or reactions. The consequence of the application of the time-reversal symmetry to these processes gives rise to that for every individual process, there is a reverse process, and in a state of equilibrium, the average rate of every process is equal to the average rate of its reverse process.

The principle of microscopic reversibility started with Boltzmann [1] who in 1872 proposed kinetics of gases as statistical ensemble of elementary collisions. The equations of physics have time-reversal symmetry. Hence, it follows that reverse collisions also obey time invariance.

This is the first example of microscopic reversibility. Boltzmann introduced the principle of detailed balance for collision meaning that each collision is equilibrated by its reverse collision [1] to achieve equilibrium ensemble. Onsager [2] mentioned that each individual reaction must balance itself when the equilibrium is reached. This means that the transition from A to B must take place as frequently as the reverse transition from B to A. Einstein [3] applied the principle of microscopic reversibility and detailed balance to develop a new branch of kinetic theory. In physics, the principle of detailed balance was used for transport processes, for collisions used for the absorption and emission of quanta.

Both Newton's equation of motion and the Schrödinger equation obey the time-reversal invariance in the absence of the macroscopic, magnetic fields. This means that if $X(t)$ is a solution of the equation, then $X(-t)$ is also a solution. Hence, X is the vector of all dynamic variables, which are all the coordinates of the particles for Newton's equation of motion and the wave functions in the configuration space for the

W. S. Gan, *Time Reversal Acoustics*,
https://doi.org/10.1007/978-981-16-3235-8_6

Schrödinger equation. However, the following situations will break the time-reversal symmetry:

1. If the dynamics depends on a pseudovector such as the magnetic field or the rotation angular speed of the rotating coordinate frame.
2. In weak interaction, the time-reversal symmetry may be broken and only the charge-parity-time (CPT) symmetry holds.

In physics, there are two main macroscopic consequences of the time reversibility of microscopic dynamics. These are the Onsager century, and the principle of detailed balance. The reciprocal relations were first discovered by Helmholtz and Thomson in the nineteenth century. Then Onsager [2] in 1931 extended it to the general theory. The reciprocal relations contain the symmetry of kinetic coefficients. Onsager [2] also discovered the connection between the reciprocal relations and detailed balance. From the equation of the law of mass action, the reciprocal relations appear in the linear approximation near equilibrium. This is a consequence of the detailed balance conditions. From the reciprocal relations, the damped oscillations in homogeneous closed systems near thermodynamic equilibria are impossible due to the spectrum of symmetric operation is real.

The concept of detailed balance was discovered by Boltzmann [1]. He invented the statistical description of the macroscopic process as an ensemble of the elementary indivisible events such as collisions. This was formalized in the Boltzmann equation. Boltzmann discovered that the time invariance of the Newtonian equation of motion leads to the detailed balance for collision. That is during equilibrium, the collisions have to be equilibrated by their reverse collisions. From this principle, Boltzmann derived the formula for entropy production and the H-theorem [1]. From this, microscopic reversibility was used to prove macroscopic irreversibility.

6.2 Onsager Reciprocal Relation and the Principle of Microscopic Reversibility

Onsager reciprocal relation [2] is also known as the thermodynamics of irreversible processes and the fourth law of thermodynamics. It shows symmetry of the components of the Onsager [2] matrix of transport coefficients. The reciprocal relations present that out of equilibrium, there is the equality of certain ratios that between forces and flows in thermodynamic systems, there exists a local equilibrium. Take an example of the fluid systems expressed in terms of pressure, temperature, and matter density. In this type of system, pressure differences give rise to matter flow from high to low pressure regimes and temperature differences will cause heat flowing from warmer to colder regions. When both temperature and pressure vary, pressure differences at constant temperature can give rise to heat flow and temperature differences at constant pressure will cause matter flow. Also, the matter flow per unit of temperature differences and the heat flow per unit of pressure differences are equal. Onsager [2] using statistical mechanics showed that this is due to the time reversibility of microscopic reversibility. The Onsager reciprocal relations are applicable to more general cases and can handle more than two thermodynamic forces at the same time. However, the principle of dynamic reversibility breaks down when magnetic fields are present.

The Rayleigh [4] reciprocity principle is a byproduct of the microscopic reversibility. The first reversibility principle for acoustics was formulated by Helmholtz in 1860 on the theory of airborne noise in pipe with open ends. Lamb [5] also worked on acoustic reciprocity for more general cases. However, it was Rayleigh in 1873 who proposed the general reciprocity theorem for sound. According to this principle, at any point *A* in a medium which is the source of source and sound waves may be generated, the resulting velocity potential at a second point *B* in the medium would be the same both in magnitude and phase, as it would have been at *A*, had *B* been the source of sound. This principle is valid even for an inhomogeneous medium and with the presence of absorption losses.

Next one considers the behaviour of a linear, inhomogeneous medium under time-reversal symmetry. There are two different conditions that can produce broken time-reversal symmetry. One is absorption losses represented by the material loss factor η. It macroscopically represents the microscopic absorption processes which recover energy from wave propagation. With an absorptive medium, macroscopically there is broken time-reversal symmetry because time-reversal wave absorption means wave amplification. Another source of broken time-reversal symmetry is the presence of the magnetic field $\overrightarrow{B}$. When charges move along circular paths, they generate magnetic fields. These magnetic fields flip their sign under time reversal. The principle of microscopic reversibility will produce

$$t_{AB}\left(\omega,\eta,\overrightarrow{B}\right)=t_{BA}\left(\omega,\eta,-\overrightarrow{B}\right) \tag{6.1}$$

in general where t_{AB} = complex field transmission coefficient between point *A* and *B*. This means that the reciprocity equation

$$t_{AB}\left(\omega,\eta,\overrightarrow{B}\right)=t_{BA}\left(\omega,\eta,\overrightarrow{B}\right) \tag{6.2}$$

does not obey when $\overrightarrow{B}\neq 0$.

The Rayleigh [4] reciprocity theorem is valid only when $\overrightarrow{B}=0$ and holds even when time-reversal symmetry is macroscopically broken ($\eta\neq 0$). This shows that the Rayleigh reciprocity theorem is equivalent to the principle of microscopic reversibility and is intimately related to time-reversed symmetry. The fact that magnetic field can break time-reversed symmetry has been shown by the broken of acoustic reciprocity in magnetoelastic crystals [6].

Other conditions to break time-reversed symmetry are the presence of nonlinear processes such as high acoustic intensities or due to time-dependent material properties such as that in a time-modulated geometry or medium.

6.3 Derivation of Reciprocal Relations

Onsager [2] considered the interaction of two or more transport processes such as heat conduction, electrical conduction, and diffusion that taking place simultaneously and using the analogy of the reciprocal relations connecting displacement and forces in thermodynamics and in mechanics. He derived a general class of reciprocal relations in irreversible processes from the assumption of microscopic reversibility. The earliest form of reciprocal relation is due to Thomson [7] which deals with thermoelectric phenomena. It is a symmetry condition for the relations connecting the forces with the velocities. Denoting J_1 as the electric current, J_2 as the heat flow, and X_1 the electromotive force driving the current, the force driving the flow of heat will be

$$X_2 = -\frac{1}{T}\operatorname{grad} T \tag{6.3}$$

where $T =$ absolute temperature.

If the current and the heat flow were completely independent, then

$$X_1 = R_1 J_1 \tag{6.4}$$

$$X_2 = R_2 J_2 \tag{6.5}$$

where $R_1 =$ electrical resistance, and $R_2 =$ heat resistance.

When the two transport processes interfere with each other, then

$$X_1 = R_{11} J_1 + R_{12} J_2 \tag{6.6}$$

$$X_2 = R_{21} J_1 + R_{22} J_2 \tag{6.7}$$

Then Thomson [7] arrived at the following symmetrical reciprocal relation:

$$R_{12} = R_{21} \tag{6.8}$$

Using thermodynamic reasoning with the following assumption: the electromotive forces produced by inequalities of temperature in a circuit of different metals, and the thermal effects of electric current circulating in it are subjected to the laws which would follow from the general principle of the thermodynamic theory of heat if there were no conduction of heat from one part of the circuit to another.

Expressing in terms of conductances, the reciprocal of resistances, one has

$$J_1 = L_{11} X_1 + L_{12} X_2 \tag{6.9}$$

$$J_2 = L_{21}X_1 + L_{22}X_2 \tag{6.10}$$

where $L_{11} = R_{22}/(R_{11}R_{22} - R_{12}R_{21})$ etc. The reciprocal relation (6.8) will become

$$L_{12} = L_{21} \tag{6.11}$$

So far, the above is concerned with heat conduction and electrical conduction. There is a third transport process known as diffusion in mixtures of gases and in solutions. From experience, any two possible transport processes are likely to interfere with each other.

For the case of three different transport processes are involved simultaneously, then the phenomenological relations can be expressed as the following forms:

$$J_1 = L_{11}X_1 + L_{12}X_2 + L_{13}X_3 \tag{6.12}$$

$$J_2 = L_{21}X_1 + L_{22}X_2 + L_{23}X_3 \tag{6.13}$$

$$J_3 = L_{31}X_1 + L_{32}X_2 + L_{33}X_3 \tag{6.14}$$

The three reciprocal relations are then

$$L_{12} = L_{21} \tag{6.15}$$

$$L_{13} = L_{31} \tag{6.16}$$

$$L_{23} = L_{32} \tag{6.17}$$

6.4 Time-Reversal Acoustics as an Application of Principle of Microscopic Reversibility

The time-reversal symmetry is an aspect of microscopic reversibility. To trace the origin of time-reversal acoustics, besides from the symmetry properties of acoustics fields [7], it can be considered as a component of the Rayleigh reciprocity principle [8] and in turn to the principle of microscopic reversibility.

References

1. Boltzmann, L.: Lectures on Gas Theory. University of California Press, Berkeley, CA (1964)
2. Onsager, L.: Reciprocal relations in irreversible processes. Phys. Rev. **37**(4), 405–426 (1931)
3. Einstein, A.: On the quantum theory of radiation. Phys. Z. **18**, 121–128 (1917)
4. Rayleigh, L. (J.W. Stutt): The Theory of Sound, vol. II, Dover reprint. Dover Publication, New York (1945)
5. Lamb, H.: On reciprocal theorems in dynamics. Proc. Lond. Math. Soc. **19**, 144–151 (1888)
6. Kittel, C.: Interaction of spin waves and ultrasonic waves in ferromagnetic crystals. Phys. Rev. **110**, 836–841 (1958)
7. Thomson, W. (Lord Kelvin): Proc. R. Soc. Edinburgh, p. 123 (1854); Collected Papers I, pp. 237–241
8. Gan W.S.: Gauge invariance approach to acoustic fields, Acoustical Imaging. ed. I. Akiyama, Springer, Netherlands, **29**, 389–394 (2007)

Chapter 7
Parity–Time Invariance

7.1 Meaning of Parity–Time (PT) Invariance

Parity–time invariance is a fuller aspect of time-reversal symmetry because time-reversal symmetry also represents the space inversion aspect which is parity or reflection invariance. So, parity–time invariance is also known as space–time reflection symmetry. Parity is spatial inversion. Parity transformation is also known as parity inversion. In three dimension, it refers to the simultaneous flip in the sign of all three spatial coordinates or a point reflection. In acoustics, PT symmetric medium can be realized by a complex refractive index obeying the condition $n(r) = n^*(-r)$.

The parity transformation changes the algebraic sign of the coordinate system. Parity is an important concept in quantum mechanics because the wave functions representing the particles behave in different ways upon transformation of the coordinate system describing them. The diagram (Fig. 7.1) illustrates a parity transformation in quantum mechanics. Parity transformation is the flip in the sign of *one* spatial coordinate. In three dimensions, it can also refer to the simultaneous flip in the sign of all three spatial coordinates (a point reflection).

The left–right symmetry is the coordinate-space realization of PT symmetry. Left–right symmetry is the coordinate representation of PT symmetry.

Thus, a right-hand coordinate system will become a left-hand coordinate system upon parity transformation. Two subsequent parity transformations will restore the coordinate system to its original form. A well-known example of parity non-conservation was that of the beta decay in weak interaction [1]. In 1998, Bender and Boettcher [2] found that the Hermitian requirement of Hamiltonian may be replaced by physically transparent condition of PT symmetry without violating any of the physical axiom of quantum mechanics. This can be illustrated as follows:

$$H = H^{PT} \tag{7.1}$$

where P = space-reflection operator or parity operator and T = time-reversal operator. The effect of P on the quantum mechanical momentum operator $\hat{p}$ and the

W. S. Gan, *Time Reversal Acoustics*,
https://doi.org/10.1007/978-981-16-3235-8_7

Fig. 7.1 Parity transformation

$$\mathrm{P}: \begin{pmatrix} x \\ y \\ z \end{pmatrix} \mapsto \begin{pmatrix} -x \\ -y \\ -z \end{pmatrix}$$

coordinate operator $\hat{x}$ is to change their signs:

$$P\hat{x}P = -\hat{x} \text{ and } P\hat{x}P = -\hat{p} \tag{7.2}$$

The effects of T on the quantum mechanical momentum operator $\hat{p}$ and the quantum mechanical coordinate operator $\hat{x}$ are to change the sign of $\hat{p}$ and leave $\hat{x}$ invariant:

$$T\hat{p}T = -\hat{p} \text{ and } T\hat{x}T = \hat{x} \tag{7.3}$$

Since both P and T are reflection operators, their square produces the unit operator:

$$P^2 = T^2 = 1 \tag{7.4}$$

Also, the P and T operators commute:

$$PT - TP = 0 \tag{7.5}$$

In terms of the P and T operators, the PT-reflected Hamiltonian H^{PT} in (7.1) can be defined as

$$H(PT) - (PT)H = 0 \tag{7.6}$$

PT symmetric Hamiltonian needs not be Hermitian.

The meaning of PT symmetry can be further illustrated by the Korteweg–de Vries (KdV) equation, a classical nonlinear acoustic equation with PT symmetry:

$$u_t + uu_x + u_{xxx} = 0 \tag{7.7}$$

To show that this equation is PT symmetric, one defines the classical parity reflection P by replacing x by $-x$. Since $u =$ velocity $==$ dx·dt, under P, u also changes sign to $-u$. The time-reversal T is the replacement of t by $-t$ since $u =$ dx/dt, under T, u also to be replaced by $-u$. It is to be noted that the KdV equation is not symmetric under P or T separately, but it is symmetric under PT transformation. There are many nonlinear wave equations with PT symmetry. KdV equation is the equation for solitons. The KdV equation is PT symmetric shows that the solitary wave has PT symmetry property. Further, new nonlinear wave equations can be generated from the KdV equation by introducing the real parameter ϵ as:

$$u_t - iu(iu_x)^{\epsilon} + u_{xxx} = 0 \tag{7.8}$$

which reduces to (7.7) with $\epsilon = 1$.

7.2 Quantum Time-Reversal Acoustics

In this section, one extends the PT symmetric acoustics into the quantum region. This is done by extending the PT symmetric Hamiltonian from real plane into the complex plane or complex phased space. Bender and Boettcher [2] in 1998 discovered the non-Hermitian quantum mechanics by replacing the mathematical axiom of Dirac Hermiticity by the physical axiom of PT (space–time reflection) symmetry. The spectra of the PT symmetric and non-Hermitian is real. Also, the energy levels of the PT symmetric non-Hermitian Hamiltonian are real and positive. To extend the PT symmetric acoustics into the quantum region, one will consider the propagation of phonons which obey the Schrödinger equation. Here, one will use the non-Hermitian or PT symmetric form of the Schrödinger equation. The Hermitian form of the Hamiltonian in the real plane or real phase space is given by

$$H = \hat{p}^2 + \hat{x}^2 \tag{7.9}$$

where p = momentum and x = distance travelled.

By extending this into the complex plane, it becomes

$$H = \hat{p}^2 + \hat{x}^2\,(i\hat{x})^{\epsilon} \tag{7.10}$$

where ϵ = real parameter. When $\epsilon > 0$, the eigenvalues of the Hamiltonian are real and positive, and when $\epsilon < 0$, they are of complex values.

This equation will govern the PT symmetry nature of the acoustic wave in the quantum region and also illustrates the PT symmetric properties of the acoustic waves in the quantum region.

7.3 PT Symmetric Acoustic Waves in the Complex Plane

By extending the Hermitian Hamiltonian given in Hamiltonian into the complex plane, one obtains the PT symmetric and non-Hermitian Hamiltonian given in (7.10).

The equation of motion of a classical particle in the complex plane is given by the Hamiltonian's equations with combination with Eq. (7.10) as

$$\frac{\mathrm{d}x}{\mathrm{d}t} = \frac{\partial H}{\partial p} = 2p \tag{7.11}$$

$$\frac{\mathrm{d}p}{\mathrm{d}t} = -\frac{\partial H}{\partial x} = i(2+\epsilon)(ix)^{1+\epsilon} \tag{7.12}$$

Combining Eqs. (7.11) and (7.12), one obtains the complex version of Newton's second law, $F = ma$, as

$$\frac{\mathrm{d}^2}{\mathrm{d}t^2}x = 2i(2+\epsilon)(ix)^{1+\epsilon} \tag{7.13}$$

When deriving the acoustic equation of motion in a freely vibrating medium, both inertial and elastic restoring forces act upon each particle. By applying the dynamical equation of mechanics to a single particle, its inertial influences only the translational particle of the particle motion. The inertial and elastic restoring forces in a freely vibrating medium are related through the translational equation of motion:

$$\nabla \cdot T = \rho \frac{\mathrm{d}^2}{\mathrm{d}t^2}x \tag{7.14}$$

Hooke's law states that

$$T_{ij} = c_{ijkl}S_{kl} \tag{7.15}$$

where T_{ij} = stress, S_{kl} = strain, and c_{ijkl} = elastic constant.

Substituting (7.15) into (7.14), one has

$$\nabla \cdot c_{ijkl}S_{kl} = \rho \frac{\mathrm{d}^2}{\mathrm{d}t^2}x \tag{7.16}$$

Extending (7.16) to the complex plane, one has

$$2\rho_i(2+\epsilon)(ix)^{1+\epsilon} = \nabla \cdot c_{ijkl}S_{kl} \tag{7.17}$$

Helmholtz equation is given by

$$\left(\nabla^2 - \frac{1}{c^2}\frac{\partial^2}{\partial t^2}\right)u(r,t) = 0 \tag{7.18}$$

$$\frac{\mathrm{d}^2}{\mathrm{d}t^2}x = 2i(2+\epsilon)(ix)^{1+\epsilon} = \frac{1}{\rho}\nabla \cdot T \tag{7.19}$$

Inhomogeneous Helmholtz equation is given by

$$\nabla^2 u(r,t) = \frac{1}{c^2}\frac{\partial^2}{\partial t^2}u(r,t) + f(x) \tag{7.20}$$

As ϵ increases from zero, the Hamiltonian is no longer Hermitian, but PT symmetry is maintained. When $\epsilon \geq 0$, PT symmetry is unbroken. When $\epsilon < 0$, PT symmetry is broken.

Here, we would extend Eq. (7.14) from real plane into the complex plane. The Helmholtz acoustic equation is given by

$$\nabla^2 x = \frac{1}{c^2}\frac{\partial^2}{\partial t^2}x \tag{7.21}$$

where c = sound velocity and r = spatial variable.

Combining (7.21) and (7.19), one has

$$\frac{1}{c^2}\frac{\partial^2}{\partial t^2}x = \frac{1}{c^2}2i(2+\epsilon)(ix)^{1+\epsilon} \tag{7.22}$$

with

$$\nabla^2 x = \frac{1}{c^2}2i(2+\epsilon)(ix)^{1+\epsilon} = \frac{1}{c^2}x\frac{1}{\rho}\nabla \cdot T \tag{7.23}$$

Equations (7.22) and (7.23) will now represent the set of acoustic equations of motion in the complex plane with PT symmetry.

7.4 Application of Time-Reversal Acoustics in the Complex Plane

This is a realization of PCT symmetry or non-Hermitian Hamiltonian in the complex plane but violation of the time-reversal symmetry in the complex plane using physical objects. This is the use of metamaterials in the complex plane with work of Zhu et al. [3]. Before them, the metamaterials are all based on the modulation of the real part of the acoustic parameter. For instance, one is dealing with negative bulk modulus, negative mass density, negative permeability, negative permittivity, and negative refraction but all in the real plain. Exploration of acoustic characteristics in a complex plain involves non-Hermitian acoustics with rich physics. Zhu et al. [3] introduced the concept of PT symmetry in acoustics by the design of acoustic metamaterials with complex parameters featuring the carefully balanced loss-and-gain region. Their design shows unique asymmetric reflection, resulting in unidirectional medium displaying unique scattering characteristics. Then, a PT symmetric cloak or a one-way cloak can be achieved by using transformation acoustics into the PT symmetric medium.

To design a PT symmetric medium in acoustics, one needs a complex bulk modulus with the condition

$$\kappa(\overrightarrow{r}) = \kappa^*(-\overrightarrow{r}) \tag{7.24}$$

and the mass density

$$\rho(\vec{r}) = \rho(-\vec{r}) \tag{7.25}$$

In general, the bulk modulus of an acoustic metamaterials should have a positive imaginary part to account for the material is lossy with inherent damping. So far the acoustic gain material with a negative imaginary part of the bulk modulus has not been found in nature.

In order for the experimental set up to represent PT symmetry, the real parts of the bulk modulus and the mass density have to be even function of position $z = 0$ and the imaginary part of the bulk modulus must be odd function. The underlying PT symmetries of the acoustic system impose a conservation relation on the acoustic wave-scattering properties of the composite structure. There is unidirectional transparency at the exceptional points of non-Hermitian acoustic systems. Such a unidirectionally transparent acoustic medium can be transformed into a PT symmetric acoustic cloak that makes an object completely invisible in specified directions.

The acoustic wave scattering properties of the composite structure of the PT symmetric acoustic system obey a generalized energy conservation relation. Outside the acoustic PT symmetric setup, the pressure field can be decomposed into right and left travelling waves. Zhu et al. [3] derived the scattering matrix $S(k)$ for the relation between the incoming and outgoing acoustic waves in a 1D two-port system as

$$\begin{pmatrix} p_f^+ \\ p_b^- \end{pmatrix} = S(k)\begin{pmatrix} p_f^- \\ p_b^+ \end{pmatrix}, \quad S(k) = \begin{pmatrix} t & r_R \\ r_L & t \end{pmatrix} \tag{7.26}$$

where p = acoustic pressure, r_L® = left-(right-) reflection coefficient, t = left(right-) transmission coefficient.

The PT symmetry of the acoustic system leads to

$$\begin{pmatrix} p_f^{-*} \\ p_b^{+*} \end{pmatrix} == S(k)\begin{pmatrix} p_f^{+*} \\ p_b^{-*} \end{pmatrix} \tag{7.27}$$

From (7.26) and (7.27), $S^*(k) = S^{-1}(k)$ and

$$r_L r_R^* == 1 - (\text{modulus of } t)^2 \tag{7.28}$$

Equation (7.28) can be expressed as $(R_L R_R)^{1/2} = /T - 1/$ where $R_{L(R)} = (\text{modulus of } r_{L(R)})^2$ and $T = (\text{modulus of } t)^2$ are a generalization of the energy conservation relation $T + R = 1$ associated with the lossless passive system. The reflection in a 1D lossless passive acoustic system is the same for opposite directions. They have also used the transfer matrix method in acoustics to calculate the phase and the amplitude of the left- and right-reflected waves and transmitted waves propagating through the PT symmetric acoustic system. It showed that for $T < 1$ the phase of the left reflection and right reflection should be equivalent.

For $T > 1$, there is a π phase difference between the two reflections. Moreover, for both cases, there is a $\pm\pi/2$ phase difference between the transmission and reflection parts.

7.5 PT Symmetric Acoustic Cloak

Due to the symmetry properties of the acoustic fields, acoustic wave equation is invariant under gauge transformation or coordinates transformation [4]. Hence, it is feasible to apply coordinate transformation to PT symmetric acoustic medium to design acoustic devices with unidirectional characteristics. Here, a 2D PT symmetric acoustic cloak with one-way invisibility or PT symmetric acoustic cloak is designed by Zhu et al. [3]. They constructed a PT symmetric periodic structure with unidirectional invisibility in the virtual space (r, θ) for which the complex modulation of material parameters takes the form of $A\exp(i\vec{\beta}\cdot\vec{r})$ where $A =$ amplitude of modulation and $/\vec{\beta}/ =$ periodicity of modulation.

In cylindrical coordinates, the mapping between virtual space (r, θ) and real space (r', θ') to produce a real acoustic cloak of cylindrical geometry can be expressed as $r = f(r')$ and $\theta' = \theta$ with $f(r') = b(r' - a)/(b - a)$ for $a \angle r' \angle b$. Here, a and b are the inner and outer radius of the acoustic cloaking shell.

Using coordinates transformation for the acoustic equations in orthogonal coordinates, the material parameters for the PT symmetric cloak in real space (r', θ') can be written as

$$\rho_r' = \rho_0(r'/r' - a) \tag{7.29}$$

$$\rho_\theta' = \rho_0(r' - a)/r \tag{7.30}$$

$$\kappa' = \left(\frac{b-a}{b}\right)^2 [r'/(r' - a)]\kappa_0[f(r', \theta')] \tag{7.31}$$

Due to the specific modulation in the form of $A\exp(i\vec{\beta}, \vec{r})$, a complex grating takes place that can provide a one-way wave vector $\vec{\beta}$. Incident plane waves with wave vector $\vec{k_1}$ can interact with the complex grating to produce a diffraction mode with spatial frequency $\vec{k_1} + \vec{\beta}$. If the phase-matching condition of $A = \vec{k_1} + \vec{\beta} - \vec{k_2} = 0$ where $\vec{k_2} =$ spatial frequency of the excited diffraction mode propagating in the surrounding medium is satisfied, then mode transitions between incident plane waves and the diffraction mode can take place. Then, $/\vec{k_1}/ = /\vec{k_2}/$. If $A \neq 0$ due to the momentum mismatch, Fig. 7.2 shows the one-way visibility of the PT symmetric cloak. Due to the satisfied phase-matching condition, $\vec{k_2} \simeq \vec{k_1} + \vec{\beta}$, shown by the inset vector diagram of Fig. 7.2a, for left incidence in virtual space,

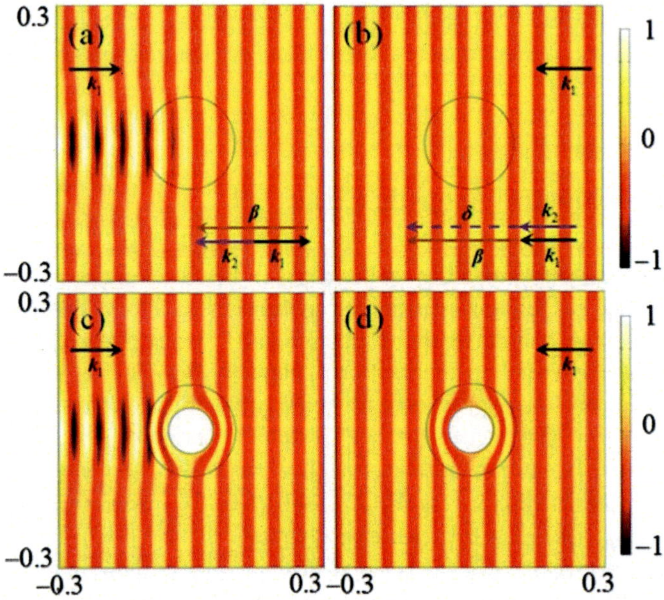

Fig. 7.2 Pressure field maps in (**a**), (**b**) the virtual space, and (**c**), (**d**) the physical space for a one-way invisible PT symmetric cloak. In (**a**) and (**c**), where the plane wave is incoming from the left, a strong backscattered reflection can be observed due to the satisfied phase matching, and therefore, the observers on the left can sense the existence of the cloak

strong Bragg reflection is observed. Due to phase mismatch for right incidence in Fig. 7.2b, scattered waves can hardly be visualized and the waves are propagating through the PT symmetric medium as if there is a bulk of surrounding medium. Figure 7.2b shows that a diffraction mode with the spatial frequency $\overrightarrow{k_1} + \overrightarrow{\beta}$ will be much larger than $\overrightarrow{k_2}$ and thus can hardly be excited since it falls into the evanescent regime. Figure 7.2c d show the corresponding simulation of the PT symmetric acoustic cloak in real space after coordinates transformation where the pressure field distribution in both incidences agree well with the results in virtual space. Here, an arbitrary object positioned in a rigid cylinder is perfectly concealed and one-way cloaked by the transformed PT symmetric medium. Also, for left incidence in Fig. 7.2c, the concealed rigid cylinder still remains invisible to the observer on the right due to the reason that the designed phase matching is only satisfied for the backward reflection. The phase-matching condition can be satisfied for the forward reflection with engineering of the direction and periodicity of the complex modulation of material parameters.

However, in (b) and (d), where the plane wave is incident from the right, the waves propagate without any scattering due to phase mismatching, and thus, an arbitrary object is perfectly concealed in the white region. The wavelengths of incident waves are $\lambda = 0.0572$ m (after Zhu et al. [3]).

References

1. Wu, C.S., Ambler, E., Hayward, R.W., Hoppes, D.D., Hudson, R.P.: Experimental test of parity conservation in beta decay. Phys. Rev. **105**(4), 1413–1415 (1957)
2. Bender, C.M., Boettcher, S.: Real spectra in non-Hermitian Hamiltonian having PT symmetry. Phys. Rev. Lett. **80**(24), 5243–5246 (1998)
3. Zhu, X., Ramezani, H., Shi, C., Zhu, J., Zhang, X.: PT-symmetric acoustics. Phys. Rev. X **4**, 031042-1–031042-7 (2014)
4. Gan, W.S.: Gauge Invariance Approach to Acoustics Fields. Springer (2019)

Chapter 8
Time-Reversal Acoustics and Superresolution

8.1 Superresolution

In a complex inhomogeneous medium, the backpropagated time-reversal sound wave will undergo multiple scatterings and multiple reflections. A small size time-reversal mirror can focus the sound wave back to the source with a spatial resolution that beats the Rayleigh resolution. This is known as superresolution. The focusing is better than sound propagation in a homogeneous medium. The sound wave is refocused on the source that emitted it. Due to the finite size of aperture of the array of transducers which is only a certain portion of the 3D time-reversal cavity, the refocusing is only approximate to describe role of evanescent wave here. This aperture size is small compared to the propagation distance, and thus, only a small portion of the advancing sound wave is captured and time reversed. In homogeneous medium, the spatial resolution of the time-reversed sound wave is proportional to the wavelength times the propagation distance and inversely proportional to the aperture size and is limited by diffraction. There is refocusing because the time-reversed sound waves propagate backwards through the medium and undergo all the refraction, reflection, and multiple scattering underwent during the forward propagation.

In an inhomogeneous medium, the time-reversed sound wave will be more sharply focused than in the homogeneous case. This is because the random inhomogeneities produce multi-path so that the time-reversed mirror (TRM) appears to have an aperture which is larger than its physical size. This is known as the effective aperture a_e. It follows that the recompressed pulse will be narrower than in the case of an homogeneous medium and produces superresolution with a spatial scale of order $\lambda L/a_e$ where L is the distance from the source to the TRM. This phenomenon was observed in experiments in the ultrasound regime [1–3] and in underwater acoustics [4–6].

Blomgren et al. [7] have demonstrated numerically superresolution imaging in underwater acoustics. They showed how multipathing in random media enhanced resolution and gave rise to superresolution. They derived an exact expression for the effective aperture of the TRM, valid both in the frequency and in the time domain for

W. S. Gan, *Time Reversal Acoustics*,
https://doi.org/10.1007/978-981-16-3235-8_8

the case when propagation distance is large compared to the sound wavelength and when both the time-reversal mirror and the correlation length of the inhomogeneities are small.

Also, in a complex inhomgeneous medium, there is a kaleidoscopic effect that creates virtual transducers that the TRM will have an effective aperture that is much larger that its physical size. Due to the multiple reverberation on the waveguide boundaries, waves emitted by each transducer are multiply reflected, creating at each reflection, virtual transducers that can be observed from the desired focal point. Thus, a large virtual array of transducers can be created from a limited number of transducers. The outcome of such operation enables creation of a kaleidoscopic transducer array from the multiplication of a small number of transducers.

Besides in a random medium made of distribution of subwavelength scatterers, a time-reversed wavefield can interact with the random media to regenerate not only propagating sound waves but also the evanescent waves or non-propagating waves which is required for refocusing below the diffraction limit.

8.2 The Effective Aperture and its Contribution to Superresolution

As mentioned in the previous section, the effective aperture plays an important role in time-reversal acoustics. The following is the derivation of the formula for the effective aperture. The effective aperture is derived from the transport limit and from systematically calculated statistics of the backpropagated phase-conjugated field using transport equation based on transport theory and Wigner distribution function in the frequency domain from the Wigner distribution function which is the Fourier transform of the Γ function and the tensor product of two Green's functions. The Wigner distribution function is a statistical probability distribution function satisfying the Wigner equation.

The Helmholtz wave equation is used, and the backpropagated wave can be expressed in terms of the Green's function. The Helmholtz equation for time-harmonic waves $u(x, y, z)\exp(-i\omega t)$ can be expressed as:

$$u_{xx} + u_{yy} + u_{zz} + k^2 n^2(x, y, z)u = 0 \tag{8.1}$$

where u = acoustic pressure, k = wave number, n = refractive index. Here, the paraxial approximation is used for the case of the time-reversal mirror with small aperture and the weak fluctuation in the propagation speed. With reciprocity, the Green's function satisfies

$$G(z, z_0; x, \xi; k) = G(z_0, z; \xi, x; k) \tag{8.2}$$

With initial source distribution at $z_0 = 0$ as $\Psi_0(\eta; k)$, the wavefield at $z = L$ is

$$\Psi(y; L; k) = \int G(L, 0; y, \eta; k)(\eta; k)d\eta \tag{8.3}$$

The time-harmonic, phase-conjugated backpropagated field at the source plane $z = o$ is given by

$$\begin{aligned}\Psi_B(\xi, L; k) &= \int G(L, 0; y, \xi; k)\overline{\Psi(y, L; k)}\chi_A(y)dy \\ &= \iint \chi_A(y)\overline{\Psi_0}(\eta; k)G(L, 0; y, \xi; k)\overline{G(L, 0; y, \eta; k)}d\eta dy \\ &= \iint \chi_A(y)\overline{\Psi_0}(\eta; k)\Gamma(L, y, y; \xi, \eta; k)d\eta dy \end{aligned}\tag{8.4}$$

where χ_A(y) = aperture function for the TRM occupying the region A and equals 1 if y ϵA and 0 otherwise and

$$\Gamma(\mathrm{L, x, y}; \xi, \eta; \mathrm{k}) = G(L, 0; x, \eta; k)\overline{G(L, 0; y, \eta; k)}$$

The Wigner distribution function can be written as

$$W(L, X, P; \xi, \eta; k) = \frac{1}{(2\pi)^d}\int \exp(iP.Y)\,\Gamma(L, X, Y; \xi, \eta; k)\,dY \tag{8.5}$$

where

$$X = (x + y)/2, Y = y - x. \tag{8.6}$$

(P, L) is the Fourier transform of $(X, L_0$ in the transverse variable x, d = number of dimensions.

For the case of a Gaussian aperture, the aperture function χ_A is a normalized isotropic Gaussian with variance a, then n

$$\chi_A(y) \sim 1\Big/\left(2\pi a^2\right)^{d/2}\exp\left[-(\text{modulus of } y)^2\ /2a^2\right] \tag{8.7}$$

Then, the backpropagated, phase-conjugated time-harmonic field from a point source will be

$$\begin{aligned}\Psi_B(\xi, L; \mathrm{k}) = \left(\frac{k}{2\pi L}\right)^d \exp\left[-a^2k^2(\text{modulus of}\,\xi)^2/2L^2\right] \\ \times\ \exp\left[(\mathrm{ik}(\mathit{modulus}\text{ of}\,\xi)^2/2\,\mathrm{L}\right]\end{aligned}\tag{8.8}$$

Blomgren et al.[7] have derived the backpropagated and mean phase-conjugated time-harmonic field as

$$\langle\Psi_B(\xi,\ L;k)\rangle = \iint\left[\left[\int W\ L, y, P; \xi, \eta; k\right)dP\right]\bar{\Psi}_0(\eta, k)\chi_A(y)dyd\eta\left(\frac{\kappa}{L}\right)^d\cdot$$

$$\times\ \ \exp\left[\left(ik(\text{modulus of }\xi)^2/2\,L\int\chi_A\left[\frac{k}{L}(\eta-\xi)\right.\right.\right. \tag{8.9}$$

where $\gamma = \pi D/8\ \Upsilon = \pi D/8 =$ phase-space diffusion coefficient.
From (8.9), they obtained

$$\langle\Psi_B(\xi, L;k)\rangle. = \left(\frac{k}{2\pi L}\right)^d \exp\left[-a^2k^2(\text{modulus of }\xi)^2/2L^2\right]k^2\xi^2\exp\left(ik\xi^2/2\,L\right) \tag{8.10}$$

Comparing with (8.8), they obtained the effective aperture for the TRM for mean time-harmonic wave as

$$a_e = \mathrm{a}\sqrt{(1+2\gamma L^3/a^{2)}} \tag{8.11}$$

8.3 Transport Theory Approach to Superresolution

In 1966, Woon Siong Gan coined and invented the name transport theory in condensed matter physics [8] during his Ph.D. works. Transport theory is today the foundation of theoretical designs of new materials. It is also a very important theory in condensed matter physics and statistical mechanics due to its relation with phase transition. Transport theory is used in the treatment of sound propagation in random media [9]. Ryzhik et al. [10] have developed a theory for the transport approximation based on Wigner distribution function. In time-reversal acoustics, it is used in the statistics of the backpropagated phase-conjugated sound field. Here, the Wigner probability distribution function given in Eq. (8.5) is used for the transport theory. The effective aperture is derived from the transport limit as well as the diffraction limit. Blomgren et al. [7] based on the Wigner distribution function have derived the transport equation as

$$k\frac{\partial W}{\partial L} + P.\nabla_x\,W = \frac{\pi k^3}{4}\hat{R}\left[(P^2-Q^2)/2k, P-Q\right]$$
$$\times\ \ [W(L, X, Q : \xi, n, k) - W(L, X, P; \xi, \eta, k)]dQ \tag{8.12}$$

where

$$W(0, X, P; \xi, \eta, k) = \exp[-iP.(\xi-\eta)/\in]/(2\pi)^d\left[\delta\left(X-\frac{\xi+\eta}{2}\right)\right] \tag{8.13}$$

with $\epsilon = \lambda/L$, $\mu = n^2 - 1$ = fluctuation of the refractive index and

$$R = \text{correlation function} = R(z, x) =< \mu(\eta, y)\mu(\eta + z, y + x) > \quad (8.14)$$

and $\hat{R}(p, P)$ is its Fourier transform in (z, x), that is, the power spectral density of the fluctuation of the refractive index.

Transport theory which is part of statistical mechanics gives a deeper analysis and understanding of sound propagation in inhomogeneous random media, and hence better understanding of the superresolution concept in time-reversal acoustics.

References

1. Derode, A., Roux, P., Fink, M.: Robust acoustic time reversal with high-order multiple scattering. P. Rev. Lett. **75** (23):4206–4209 (1995)
2. Fink, M.: Time reversed acoustics. Phys. Today **50**, 34–40 (1997)
3. Fink, M.: Time-reversed acoustics. Sci. Am. 91–97 (1999)
4. Dowling, P.R., Jackson, D.R.: Phase conjugation in underwater acoustics. J. Acoust. Soc. Am. **89**, 171–181 (1990)
5. Hodgkiss, W.S., Song, H.C., Kuperman, W.A., Kal, T., Ferla, C., Jackson, D.R.: A long-range and variable focus phase-conjugation experiment in shallow water. J. Acoust. Soc. Am. **105**, 1597–1604 (1999)
6. Kuperman, W.A., Hodgkiss, W., Song, H.C., Akal, T., Ferla, C., Jackson, D.R.: Phase conjugation in the ocean. J. Acoust. So. Am. **102**, 1–16 (1997)
7. Blomgren, P., Papanicolaou, G., Zhao, H.: Superresolution in time-reversal acoustics. J. Acoust. Soc. Am. **111**, 230–248 (2002)
8. Gan, W.S.: Transport Theory in Magnetoacoustics, PhD thesis. Imperial College, London (1969)
9. Van Rossum, M.C.W., Nieuwenhuizen, T.M.: Multiple scattering of classical waves:microscopy, mesoscopy, and diffusion, and diffusion. Rev. Mod. Phys. **71**(1), 313–371 (1999)
10. Ryzhik, L.V., Papanicolaou, G.C., Kellen, J.R.: Transport equation for elastic and other waves in random media. Wave Motion **24**, 327–370 (1996)

Chapter 9
Spontaneous Time Reversal Symmetry Breaking

9.1 Introduction

This chapter is about spontaneous time reversal symmetry breaking. This is an outcome of irreversibility of the system [1]. For a closed system, this depends on the initial solution of the equation of motion which breaks the time reversal symmetry. For an open system, the initial condition of the environment has to be included in the effective dynamics. This can be done by the generalization of the ϵ-prescription when the spontaneous breaking of the time-reversal symmetry will occur with the nonuniform convergence of the limit $\epsilon \rightarrow 0$. Here, one will start with the closed system obeying time-reversal invariance. The dynamics of the system can be broken by a mechanism similar to that of spontaneous symmetry breaking. The time-reversal symmetry can be broken by the boundary conditions. The spontaneous breakdown of the time-reversal symmetry is part of the time arrow problem [2, 3]. This is the emergence of an orientation for the time in a time-reversal invariance microscopic dynamics. The second law of thermodynamics with the non-decreasing nature of entropy for closed system is an explanation of the directed flow of time in macroscopic phenomenon. The entropy production and the thermodynamics time arrow are related through the dissipative work, in terms of the relative entropy of a driven system and its time reversed form [4].

The initial conditions play a role in the breaking of time reversal symmetry even if the equation of motion is symmetric. Irreversibility plays an important role in spontaneous broken time-reversal symmetry. In the following section, one will construct an order parameter for the time-reversal symmetry with the use of the extended action principle. Here, the time arrow, defined by the sign of the order parameter, is mechanical in nature since it does not refer to a thermal bath or an equilibrium state. It refers to an ordered initial condition, and so, its direction should agree with the thermodynamical time arrow.

The time arrow is a concept positing one-way direction or asymmetry of time. Physical processes at the microscopic level are entirely or mostly time symmetric. That is, if the direction of time were to reverse, the equations would remain

W. S. Gan, *Time Reversal Acoustics*,
https://doi.org/10.1007/978-981-16-3235-8_9

unchanged. However, at the macroscopic level, this is often not the case, meaning there is a direction or flow of time.

9.2 Determination of the Spontaneous Broken Time Reversal Symmetry for the Case of the Harmonic Oscillator

Here, one will first introduce the generalized action principle [5]. The variational principle or the action principle will be used. The problem is to find the dynamics of a system, represented by a coordinate x interacting with the environment described by the coordinates $y = (y_1, \ldots y_N)$. It is assumed that the full system, described by the coordinates *x*, *y*, follows a time reversible and conservative dynamics which keep the coordinates bounded at finite energy. The corresponding action can be written as

$$S[x, y] = S_s[x] + S_e[x, y] \tag{9.1}$$

In order to have unique solution of the equation of motion, auxiliary conditions are needed. These are initial conditions for the system, and the environment at the initial time $t = t_i$, and the system trajectory is followed until a final time $t = t_f$. The equation of motion to be followed by the system trajectory is derived in two steps. First, the environment equation of motion is solved for a general system trajectory, $x(t)$.

Then, the solution $y[t; x]$ is substituted into the system equation of motion. The generalization of the action principle can handle initial condition problems as opposed to the traditional variation method which depends on boundary conditions. The treatment of the problem will be as follows. First, the system follows its time evolution from the initial time t_i until the final time t_f. Then, a time inversion will take place on the state of motion and the reversed motion will be followed until the original initial conditions are reversed.

The trajectory can be represented as a closed loop as

$$\tilde{x}(t) = \mathrm{x(t)} \quad t_i < \quad \mathrm{t} \quad > t_f \tag{9.2}$$

$$= \; x\left(2t_f \; - \; t\right) \qquad t_f < \mathrm{t} < 2t_f \; - \; t_i \tag{9.3}$$

The action for this closed loop can be given as

$$S[\tilde{x}] \; = \; \int_{t_i}^{t_f} dt L\left[\tilde{x}\,(t), \dot{\tilde{x}}\,(t)\right] \; + \; \int_{t_f}^{2t_f - t_i} dt L^T\left[\tilde{x}\,(t), \dot{\tilde{x}}\,(t)\right] \tag{9.4}$$

where T denotes time reversal, $\tilde{x}^T(t) \; = \; \tilde{x}\;(2t_f \; - \; t_i \; - t)$, and $L =$ Lagrangian.

The trajectory $\tilde{x}(t)$ can be split into two segments,

$$\widehat{x(t)} = (x^{+}(t), x^{-}(t)) = (\tilde{x}(t), \tilde{x}(2t_f - t) \quad (9.5)$$

Then, (9.4) can be written as

$$S[\tilde{x}] = \int_{t_i}^{t_f} dt\big[L(x^{+}(t), \dot{x}^{+}(t)\big) - L\big(x^{-}(t), \dot{x}^{-}(t)\big] \quad (9.6)$$

Both x^{+}(t) and x^{-}(t) satisfy the same initial conditions. The doublet $\hat{x} = (x^{+}, x^{-})$ will be known as chronon because it represents the direction of the flow of time within the action principle, and the time reversal transformation is known as chronon conjugation. Then, a physical trajectory satisfying the equation of motion is chronon conjugation invariant

$$x^{+}(t) = x^{-}(t) \quad (9.7)$$

The action of the extended trajectory Eq. (9.5) is degenerated within the space of chronon conjugation invariant trajectories. With the extension of the action to handle the initial condition, one needs a non-degenerate action functional. This requires adding an infinitesimal piece to the closed time path Lagrangian Eq. (9.7):

$$S[\tilde{x}] = \int_{t_i}^{t_f} dt\big[L(x^{+}, \dot{x}^{+}) - L(x^{-}, \dot{x}^{-})\big] + S_{\epsilon}[x] \quad (9.8)$$

where

$$S_{\epsilon}[\hat{x}] = \mathrm{i}\frac{\varepsilon}{2}\int_{t_i}^{t_f} dt\big(x^{+2} + x^{-2}\big) \quad (9.9)$$

The advantage of removing the degeneracies with an imaginary term is that time reversal can be represented by flipping the sign of the real part of the action:

$$S^{T}[\hat{x}] = S[\tau\hat{x}] = -S^{*}[\hat{x}] \quad (9.10)$$

The time reversal can be achieved also by performing these transformations: $t \rightarrow 2t_f - t_i - t$ and $\in \rightarrow -\in$.

References

1. Prigogine, I.: Irreversibility as a symmetry-breaking process. Nature **246**, 67–71 (1973)
2. Halliwell, J.J., Pėrez-Mercador, J., Zurek, W.H.: Physical Origin of Time Asymmetry. Cambridge University Press, Cambridge, UK (1994)

3. Savitt, S.: Time' Arrow Today. Cambridge University Press, Cambridge, UK (1994)
4. Crooks, G.E.: Non equilibrium measurements if free energy differences for microscopically reversible Markovian system. J. Stat. Phys. **90**, 1481–1487 (1998)
5. Polonyi, J.: Environmental induced time arrow, arxiv: 1206,5781 (2012)

Chapter 10
Quantum Time-Reversal Symmetry

10.1 Introduction

The classical time-reversal acoustics is a well-established field in acoustics. It is based on the time-reversal symmetry property of the acoustic field and the acoustics equation of motion [1] and has found several important applications in technologies.

It is known that the Schroédinger equation of quantum mechanics also has time-reversal symmetry property, and also, there is time-reversal symmetry in interaction in particle physics. These are examples of quantum time-reversal symmetry. In this chapter, the properties of quantum time-reversal symmetry will be studied and its applications in quantum acoustic metamaterials will be given.

10.2 Time-Reversal Operator and Its Properties

In classical, time--reversal symmetry means the replacement of t in the equation by $-t$ will leave the equation unchanged. For instance, consider Newton's second law of motion:

$$F = \mathrm{ma} = m\frac{\mathrm{d}^2}{\mathrm{d}t^2}t \tag{10.1}$$

Replacing t by $-t$, then

$$F = m\frac{d^2}{(-t)^2} = m\frac{\mathrm{d}^2}{\mathrm{d}t^2}t \tag{10.2}$$

which leaves the equation unchanged.

However, this does not apply to the quantum case. The Schroédinger equation is not time-reversal invariant in this case as can be shown below:

W. S. Gan, *Time Reversal Acoustics*,
https://doi.org/10.1007/978-981-16-3235-8_10

$$\lambda(\overrightarrow{r})\nabla\left[\frac{1}{\rho(\bar{r})}\nabla p(\overrightarrow{r})\right]+\omega^2 p(\overrightarrow{r})=0$$

$$H\Psi = i\hbar\frac{\partial\Psi}{\partial t} \tag{10.3}$$

where H = Hamiltonian, Ψ = wave function, and $\hbar$ = Planck's constant.

By replacing t by $-t$, one has

$$H\Psi = i\hbar\frac{\partial\Psi}{-\partial t} = -i\hbar\frac{\partial\Psi}{\partial t} \tag{10.4}$$

instead.

Quantum time-reversal symmetry can be implemented in the following way: if H is real and time independent, then if $\Psi(x, t)$ satisfies the Schroédinger equation, then $\Psi^*(x, -t)$ also satisfies the equation where Ψ^* = complex conjugate of Ψ.

From (10.3), $\mathrm{H}\Psi(\mathrm{x}, \mathrm{t}) = \mathrm{i}\hbar\frac{\partial\Psi(x,t)}{\partial t}$, then

$$(\mathrm{H}\Psi(\mathrm{x}, \mathrm{t}))^* = \left(\mathrm{i}\hbar\frac{\partial\Psi(x, t)}{\partial t}\right)^* \tag{10.5}$$

If H is real, then $\mathrm{H}\Psi^*(x, t) = -i\hbar\frac{\partial}{\partial t}\Psi^*(x, t)$.

Replacing t by $-t$, then

$$\begin{aligned} &\mathrm{H}\Psi^*(\mathrm{x}, -\mathrm{t}) = -\mathrm{i}\hbar\Psi^*(\mathrm{x}, -\mathrm{t})/-\partial\mathrm{t} \\ \text{or}\quad &\mathrm{H}\Psi^*(\mathrm{x}, -\mathrm{t}) = \mathrm{i}\hbar\Psi^*(\mathrm{x}, -\mathrm{t})/\partial\mathrm{t} \end{aligned} \tag{10.6}$$

Comparing (10.3) and (10.6), it means

$$\Psi(x, t) = \Psi^*(x, -t) \tag{10.7}$$

Another property of quantum time-reversal symmetry can be shown below.

The Schroédinger equation is given by time-reversal symmetric if

$$\mathrm{i}\hbar\frac{d}{dt}/\Psi >= \mathrm{H}/\Psi > \tag{10.8}$$

It is time-reversed symmetric if the Hamiltonian H commutes with the time-reversed operator T.

In quantum time-reversal symmetry, the time-reversal operator consists of two combined operations of negating the time t and the complex conjugation. That is

$$t > -t \text{ and } i \quad > -I \tag{10.9}$$

Using the commutability relation

$$[H \cdot T] = 0 \tag{10.10}$$

And applying the time-reversal operator T from the left of the Schroédinger Eq. (10.8), one will obtain

$$i\hbar \frac{\mathrm{d}}{\mathrm{d}t}(T/\Psi >) = H(T/\Psi >) \tag{10.11}$$

This shows that the time-reversed state $t/\Psi >$ satisfies the same equation as the original state $/\Psi >$.

Also, Eugene Wigner has defined quantum time-reversal symmetry by stating that the time-reversal operator is anti-unitary. That is, it maps i to $-i$ which is given in (10.9) above. From this, then the action of time reversal on a wave function will contain two parts: first perform complex conjugation in a certain basis followed by a linear transformation in this basis. That is

$$I = \mathrm{UK} \tag{10.12}$$

where U represents unitary transformation and K represents complex conjugation. This can be further illustrated as the time reversal acts on the operators as

$$\mathrm{TOT}^{-1} = \mathrm{UKOKU}^{+} = \mathrm{UO}^{*}U^{+} \tag{10.13}$$

That is, the action of time reversal on operators consists of two steps: first take complex conjugation of the operator written in certain basis, then followed by the conjugation of the operator by some unitary transformation in time basis. This shows the anti-unitary property of time reversal. Anti-unitary also implies that.

1. Time reversal does not commute with complex numbers or

$$\mathrm{TCT}^{-1} = C^{*} \tag{10.14}$$

2. Time reversal is not a linear operator on wave functions or

$$T(\alpha_1\Psi_1 + \alpha_2\Psi_2) = \alpha_1^{*}T(\Psi_1) + \alpha_2^{*}T(\Psi_2) \tag{10.15}$$

   That is, it is anti-linear. Anti-linear operators are allowed as symmetry operators in quantum mechanics.
3. The absolute value of wave function inner product remains invariant. This can be shown as follows:

$$T\Psi_1 = U\Psi_1^{*}, \quad T\Psi_2 = U\Psi_2^{*} \tag{10.16}$$

Therefore,

$$\langle T\Psi_1/T\Psi_2\rangle = \langle \Psi_1^{*}/U^{+}U/\Psi_2^{*}\rangle = \langle \Psi_1/\Psi_2\rangle^{*} \tag{10.17}$$

Eugene Wigner showed that in order for the absolute value of inner product to be invariant, the symmetry operation must either be unitary or anti-unitary.

10.3 Application of Quantum Time-Reversal Symmetry

Acoustic metamaterials can be applied to time-reversal acoustics as can be shown in the description below. Acoustic metamaterial can convert evanescent waves into propagating waves and enables the capture of information given by the evanescent waves to achieve sub-wavelength image resolution limit at the far field. In this sense, acoustic metamaterial can be applied to time-reversal mirror. First one fabricates the time-reversal mirror using acoustic metamaterials. With the time-reversal symmetry property of the acoustic wave equation of motion, one is able to focus back into the initial source position from the far field. With the time-reversal symmetry property between the propagating wave and the evanescent wave, one is able to obtain a sub-wavelength resolution at the focal spot. The availability of the sub-wavelength information in the free space radiation is due to the resonance enhancement of the eigenmode of the broadband source placed at the near field of the medium. This occurs during the conversion of the evanescent wave into propagating wave by the acoustic metamaterial. This gives rise to sub-wavelength image resolution.

Lemoult et al.[2] have experimentally demonstrated the use of acoustic metamaterial in time-reversal acoustics. The unit cells of the acoustic metamaterial are Helmholtz resonators. In the far field, an array of soda cans is used. The concept of the resonant metamaterial is used. The experiment demonstrates that the array of cans excited in resonant periodic modes in sub-wavelength spatial period. Commercial computer loudspeaker is used to generate incident waves which are diffraction limited. It is found that these sub-wavelength resonant modes are Bloch modes and with radiation patterns depending on their wave vectors. From this sub-wavelength focusing of sub-diffraction wave field of broadband sound can be obtained in the far field. They are able to focus on sound into spots as small as 1/25th of sound wavelength in the air and a position resolution of the centre to centre distance between the soda cans of up to 1/15th of sound wavelength. The experimental setup is shown in Figs. 10.1 and 10.2 shows the array of cans surrounded by loudspeakers.

Their experiment showed that incident monochromatic and diffraction limited acoustic fields can generate sub-diffraction resonant modes over the whole array of soda cans and produce sub-wavelength focal spots and time-reversal focusing sound from an array of sensors into the point. This amounts to the summing of all modes with zero relative phase at this specific position. This creates a spatiotemporally focused wave. This can be extended to a broadband source by adding up various presenting modes coherently at a given location with specific time while adding up incoherently and other positions and other times.

The experiments have been performed at several locations using eight computer loudspeakers with the ensemble of Helmholtz resonators which are the soda cans. It is found that much thinner foci can be achieved with the array of cans than with

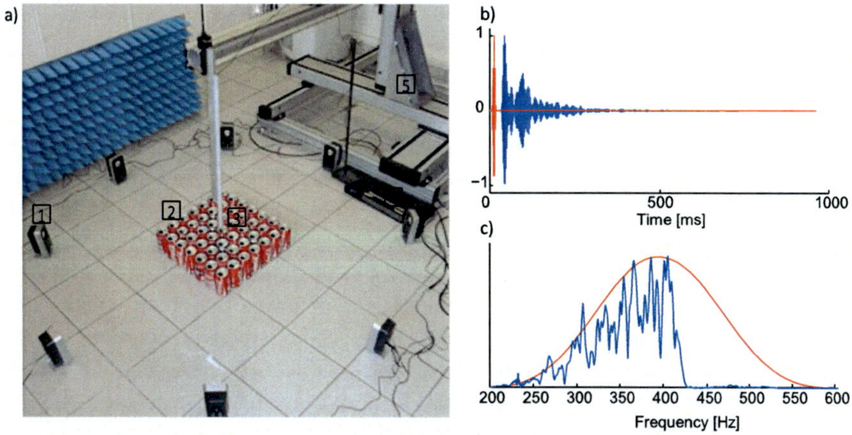

Fig. 10.1 **a** Experimental setup detailed in the text. **b** typical emitted pulse (red) and measured pressure (blue) on top of one can. **c** spectra of (**b**). (After Lemoult et al. [2])

Fig. 10.2 Close view of the array of scans surrounded by the eight computer speakers. (After Lemoult et al. [2])

a single can. They also discovered that the sub-wavelength focusing of acoustic pressure fields can lead to strong enhancement of the acoustic displacements. This important property can be used for the design of high performance actuators and sensors. It also has the advantage of dispersion which enables one to independently address many sensors with a few sources and with their temporal signatures. Besides this, it possesses the capability of engineering a matrix of actuators and sensors arranged on a sub-wavelength scale.

Lemoult et al.[2]'s results can be extended to a broad frequency spectrum of resonators and generalized to elastic waves propagation in solids. This will have several potential applications such as the design of micromechanical actuators at any frequency, arrays of actuators and audible sound control for audible sound generation.

Acoustic metamaterials can enhance image resolutions in time-reversal acoustics application to acoustical imaging. This is known as superresolution. This enables the projection on to far field the near field information of a source regardless of its spatial variation.

It is of interest to extend from the classical regime to the quantum regime with the application of quantum acoustic metamaterials to quantum time-reversal acoustics. Quantum time-reversal acoustics is a new field which is an application of the quantum time-reversal symmetry to acoustics. The acoustic metamaterials which can be applied to classical time-reversal acoustics in this case will be extension to the application of quantum acoustic metamaterials to quantum time-reversal acoustics. The advantage of quantum acoustic metamaterials compared with classical acoustic metamaterials is higher sensitivity. In this new subject, three aspects have to be considered: one is the use of the quantum form of the split ring resonator (SRR) as the unit cell in quantum acoustic metamaterials. One has to shrink the size of the SRR to microscopic size. We start with a SRR used by Movchan et al. [3]. We then shrink the size of the SRR to microscopic size. Secondly, one has to consider the conversion of the evanescent waves to propagating waves in the quantum acoustic metamaterials. Thirdly, the taking place of local resonance in quantum acoustic metamaterials as in classical acoustic metamaterials. In 2002[3], SSR was introduced as the unit cell of acoustic metamaterials. It is necessary to have sonic crystals to have an analysis of the frequency band gap characteristics as these are derived from the inherent limiting properties of artificial SRR. This is a description of the mechanical properties for sonic crystals as a macroscopically homogeneous substance. Hence, we have below a description of the theory of quantum acoustic metamaterial based on band gap study. The study of sonic crystal involves the calculation of acoustic wave propagation in periodic structures. The ultimate purpose is to find the existence of band gap in the periodic structure. The multiple scattering (MS) approach of Kafesaki and Economous [4] is used. This is based on the well-known Korringa–Kohn–Rostoker (KKR) theory [5, 6]. They start with the acoustic wave equation in a periodic medium:

$$\lambda(\vec{r})\nabla\left[\frac{1}{\rho(\vec{r})}\right]\nabla p(\vec{r}) + \omega^2 p(\vec{r}) = 0 \tag{10.18}$$

where p = acoustic pressure, $\rho(\vec{r})$ = mass density, ω = sound angular frequency, $\lambda = \rho\left(c_l^2 - c_t^2\right)$ = Lamé coefficient of the medium, c_l, c_t = longitudinal and transverse sound velocity, respectively.

Equation (10.18) can be rewritten as:

$$\begin{aligned}&\nabla^2 p(\vec{r}) + \frac{\omega^2}{c_0} p(\vec{r}) + \omega^2\left[\frac{1}{c^2(\vec{r})} - \frac{1}{c_0^2}\right]p(\vec{r}) + \\ &\rho(\vec{r})\left[\nabla\frac{1}{\rho(\bar{r})}\right]\nabla p(\vec{r}) = 0\end{aligned} \tag{10.19}$$

which as the form

$$H_0(\vec{r})p(\vec{r}) + U(\vec{r})p(\vec{r}) = 0 \quad (10.20)$$

where $H_0(\vec{r})p(\vec{r}) = 0$, $H_0(\vec{r}) = \nabla^2 + \frac{\omega^2}{c_0^2}$ represents the wave equation in the absence of scatterer and c_0 is the wave velocity in the host metal.

Equation (10.20) has the same form as the Schroédinger equation for the electron waves. This analogy shows that one can extend the KKR to the acoustic case. However, one has to beware of the important difference between the electron case and the acoustic case. That is the potential in the acoustic case has a δ function singularity at the surface of the scatterers due to the factor ∇p^{-1}. Thus, the contribution of the surface scattering to the volume integral is not negligible as in the electron case [5].

It is shown by [6] that in a periodic system, Eq. (10.19) can be written as:

$$P(\vec{r}) = \int_U G\left(\vec{r} - \vec{r'}\right) V(\vec{r}) p(\vec{r}) d\vec{r'} \quad (10.21)$$

where $V(\vec{r})$ = local potential.

The Green's function $G\left(\vec{r} - \vec{r'}\right)$ is given by

$$G\left(\vec{r} - \vec{r'}\right) = \sum_n \exp\left(i.\vec{k} \cdot \vec{R_n}\right) G_0\left(\vec{r} - \vec{r'} - \vec{R_n}\right) \quad (10.22)$$

where G_0 is the Green's function for the homogeneous equation $H_0(\vec{r})\mathrm{p}(\vec{r}) = 0$:

$$\mathrm{G}\left(\vec{r} - \vec{r'}\right) = -\frac{1}{4\pi}\left[\exp\left(ik_0/\vec{r} - \vec{r'}\right)//\vec{r} - \vec{r'}/\right] \text{ and } k_0 = \omega/c_0 \quad (10.23)$$

$V(\vec{r})$ is zero outside the unit cell centred at the origin of the coordinate system. It is related to U by $U(\vec{r}) = \sum_n V\left(\vec{r'} - \vec{R_n}\right)$ and the pressure field, $p(\vec{r})$, obeying the Bloch 's condition, $p\left(\vec{r} + \vec{R_n}\right) = \exp\left(i\,\vec{k} \cdot \vec{R_n}\right) p(\vec{r})$.

Taking into account that for acoustic waves, the local potential V is nonzero only inside, and at the surface of the scatterers, Kafesaki et al.[4] obtained the multiple scattering equation as:

$$\sum_{l',m,}^{\prime} \left[A_{lml'm'} \quad - k_0 I_m\left(t_{l'}^{-1}\right)\delta_{ll}, \delta_{mm'}\right] a_{l'm,} = 0 \quad (10.24)$$

where $A_{lml'm'}$, $\delta_{mm'}$ and $a_{l'm'}$ = structure constants, t_l = scattering coefficients. The above equations show that the elastic parameters of the scattering material affect the calculations only through the scattering coefficients t_l which can be calculated very easily and accurately for both solid and fluid scatterers. Thus, this method can be applied to both solid and fluid scatterers changing only the form of a single scattering problem.

References

1. Gan, W.S.: Gauge invariance approach to acoustic fields. In Proceedings of ICSV14. Cairns, Australia (2007)
2. Lemoult, F., Fink, M., Lerosey, G.: Acoustic resonators for far field control of sound on a subwavelength scale. Phy. Rev. Lett. **107**, 064301 (2011)
3. Movchan, A.B., Guenneau, S.: Split-ring resonators and localized modes. P. Rev. B, **70**(12), 125116 (2004)
4. Kafesaki, M., Economou, M.: Multiple scattering theory for three-dimensional periodic acoustic composites. Phy. Rev. B **60**, 11993–12001 (1999)
5. Korringa, J.: On the calculation of the energy of a Bloch wave in metal, Phys. (Amsterdam) **XIII**, 392 (1947)
6. Kohn, W., Rostoker, N.: Solution of the Schroédinger equation in periodic lattices with application to metallic lithium. Phy. Rev. **94**, 11111 (1951)

Chapter 11
Topology in Acoustics and Topological Sound Waves

11.1 Significance of Topology in Physics and Topological Sound Waves

Topological sound waves are related to non-reciprocal sound waves. For instance, quantum Hall insulator breaking time-reversal symmetry and topological edge states are based on one-way sound propagation and have been demonstrated in several types of photonic crystals. Symmetry and topology are concepts central to the understanding of physics. Symmetry plays a key role in phase transition such as the Ginzburg–Landau theory [1] of spontaneous symmetry breaking of second-order phase transition. Different phases of matter are characterized by their underlying symmetry which are spontaneously broken. For instance, by the breaking of the translational symmetry, the solid phase crystal has become the liquid phase water. With the discovery of the topological phase transition [2] and topological insulator, it is known that topology also plays a role in phase transition. In 1980, von Klitzing discovered quantum Hall effect and the first observation of the topological phases of matter. It was found that a two-dimensional electron gas sample, subjected to strong magnetic field and low temperature, produces a quantized Hall conductance, immune to impurities and independent of sample size. It was also demonstrated that the state of the matter responsible for such phenomena is characterized by a completely different classification paradigms based on the concept of topological order [3] which describes different phases of matter beyond symmetry breaking. This means that two different phases of matter can have the same symmetry. This is a topological description of phases of matter based on topological orders. This is beyond the notion of symmetry breaking. This concerns the fundamental properties of the system that are insensitive to continuous perturbation of material parameters. They change only under quantum phase transition. In quantum Hall effect, the Hall conductance is a fundamental property. The non-trivial topological properties of the energy bands give rise to its quantization. The topological properties are featured with a nonzero topological invariant, the Chern number [3]. The Chern number characterizes the geometric phase or the Berry phase [4] accumulation over the Brillouin zone.

W. S. Gan, *Time Reversal Acoustics*,
https://doi.org/10.1007/978-981-16-3235-8_11

Hence, it is closely related with the behaviour of the energy bands in the momentum space.

It is known that the time-reversal symmetry will be broken by a periodic magnetic field and produces nonzero Chern number [5]. A topological insulator will occur. This has the electronic property that is conducting on the edge but insulating in the bulk. This non-trivial topological system supports a gapless edge state in the bulk energy gap. It is essentially unlike a normal insulator which has zero Chern number [5].

Besides the effect of adding a magnetic field, it is also known that non-trivial topological phases [6, 7] can be generated by inherent spin–orbital coupling of a material. It was also found that a pair of gapless edge states emerges in the insulating bandgap in system with spin–orbital coupling. The edge states with conjugate electron spins exhibit spin-dependent propagation behaviour. This is known as the quantum spin Hall effect (QSHE). Here, the total Hall conductance is zero with the time-reversal symmetry intact as the Chern number is zero.

Besides this, another discrete degree of freedom, the valley, has been proposed to realize a topological state, the quantum valley Hall effect (QVHE) [8]. Valley represents the two energy extrema of the band structure in momentum space with the Berry curvature exhibiting opposite signs and integral over the full Brillouin zone zero while the integral within each valley as nonzero. This shows a valley-selective topologically non-trivial property. The QVHE also obeys time-reversal symmetry. The above are topological phases which share the same property that the edge states span the bulk band gap and separate domains with different Chern numbers. The above topological states have potential applications in topological quantum computing and the next generation of electron devices. However, the realization of these topological phases in electron systems pose great challenges such as the validity of the single electron approximation and the inevitable material defects. Hence, several quantum topological states have been extended to photons and photonic systems which have large scale in both space and time. This enables easier fabrication and more accurate measurement and emulates topological phases to the classical region. However, breaking the time -reversal symmetry in the photonic and phononic systems requires additional efforts to break time-reversal symmetry compared with the electronic system due to photons, and phonons do not carry a half-integer spin, and so, cannot interact with the magnetic field directly. On top of this, there is substantial difference between photonic system and phononic system because longitudinal sound waves lack a transverse polarization degree of freedom which can be used to construct pseudospins to mimic the up and down spin in electronic system. Hence, there are additional complexities to implement the quantum effects.

11.2 Acoustical Quantum Hall Effect

The acoustical quantum Hall effect was introduced by Fleury et al. [9] using moving air flow in ring cavities to break time-reversal symmetry. The acoustic non-reciprocity is due to the imparted airflow having the role of magnitude bias, splits the degenerary of the two counter-propagating azimuthal resonant modes in the ring cavities. Consider the following equation describing sound propagation in a circulating airflow with velocity $\overrightarrow{v}$:

$$\left[\left(\nabla - i\overrightarrow{A_{\text{eff}}}\right)^2 + \omega^2/c^2 + \left(\frac{\nabla\rho}{2\rho}\right)^2 - \frac{\nabla^2}{2\rho}\ \rho\right]\phi = 0 \tag{11.1}$$

where ϕ = vector potential, ω = angular frequency, c = sound speed, ρ = mass density of air.

For nonzero $\vec{v}$, $\overrightarrow{A_{\text{eff}}} = -\omega\ \vec{v}/c^2$ = effective vector potential which generates the effective magnetic field $= \overrightarrow{B_{eff}} = \nabla \times \overrightarrow{A_{eff}}$ which breaks the time-reversal symmetry.

With the above principle of the acoustic quantum Hall effect, various designs of phononic crystals have been fabricated. One of them is given in [10]. Here, a hexagonal lattice and a honeycomb lattice are designed. A pair of Dirac-like points deterministically appear at the Brillouin zone boundary for $\vec{v} = 0$ due to the intrinsic lattice symmetry. Also, the time-reversed symmetry is broken when the airflow is introduced. This is illustrated by a bandgap opening shown in Fig. 11.1. With the Chern number with nonzero values, the systems are in the topologically non-trivial phase. This implies the presence of one-way edge states along its boundaries when truncated by other topologically trivial system with Chern number = 0 and the unidirectional behaviour of acoustic wave propagation, protected by topologies and robust against various sharp bends and defects.

The above airflow-based designs can be experimentally implemented by uniformly biased circulators with challenges such as flow instability and nonsynchronous rotation. An example is the design of Zhu et al. [11] using a rotating chiral structure based on ring resonators. This support high-order whispering gallery modes with high Q factor. Another design is the use of active liquids flowing spontaneously even without an external drive [12]. This helps to relax the experimental requirements and extend topological phases to active materials.

11.3 Acoustical Quantum Spin Hall Effect

There is also topological phase of matter observing time-reversal symmetry. This is the quantum spin Hall effect (QSHE) also known as topological insulator (13). QSHE is due to the coupling of two quantum Hall states. The coupling between spin and orbital angular momentum or the spin–orbit coupling plays a key role in QSHE. This inserts a spin-dependent force on the moving electrons even in the absence of the magnetic field. The outcome is that the electrons with opposite spin angular momentum will move in opposite directions along the edges. This phenomenon is characterized by the topological order known as the spin Chern number.

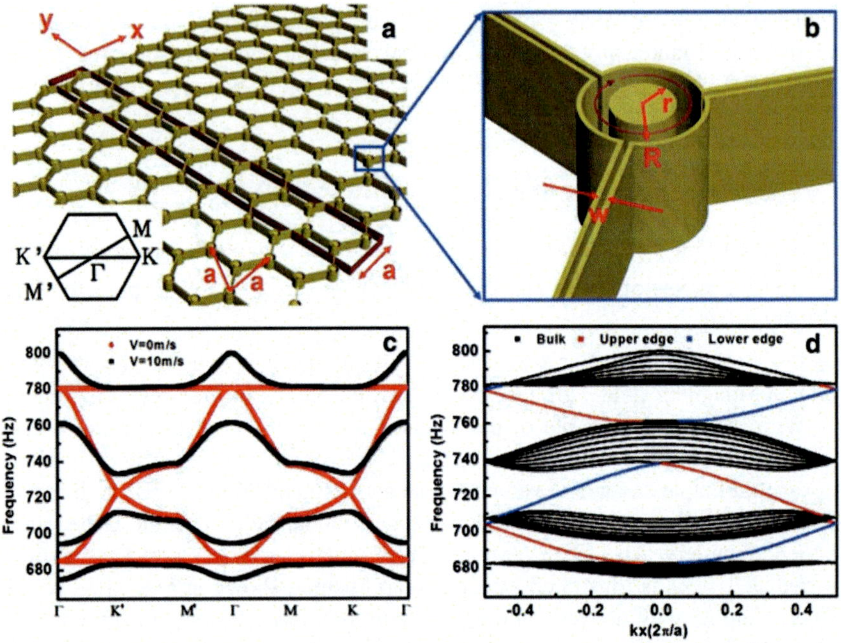

Fig. 11.1 Topological sonic crystal made of air-flow-circulated units. **a** Illustration of the sonic crystal consisting of the air-flow circulated ring resonators. The super cell (highlighted by the red rectangle) is selected to calculate the projected frequency band along the zigzag edge, and the inset shows the first Brillouin zone. **b** The zoomed-in illustration of the detailed structural parameters. **c** The bulk frequency bands of the sonic crystal without air flow ($V = 0$, denoted by red circles) and with clockwise air flow ($V = 10$ m s − 1, denoted by black squares). **d** The projected frequency bands of the sonic crystal with clockwise air flow ($V = 10$ m s − 1) along x-direction (zigzag edge). See appendix A for the case of smaller V(V = 3 ms^{-1}). (After Ni et al. [10])

There is also the design of acoustic QSHE using phononic crystals [14]. This is based on the use of coupled resonators that support clockwise and anticlockwise resonant modes that impart the pseudospin. He et al. [15] demonstrated the acoustic QSHE using a honeycomb lattice consisting of steel rods in air and based on the accidental degeneracy. This phononic crystal structure has a 2D irreducible representation of the C_{6V} symmetry and supports two pairs of degeneracies at the Γ point which can be hybridized to emulate the pseudospin. With the decrease in the filling ratio of the steel rods, one can enable the move in frequency and exchange their positions of the two pairs of dipolar and quadrupolar modes, separated by a bandgap. The doubly degenerate Dirac-like cones is formed at a point where the band gap is closed and the two pairs accidentally touch together.

This leads to a topological transition from the ordinary or trivial state to the topological or non-trivial state. This transition point is the double Dirac-like point. A pair of edge states appear in the non-trivial topological state. They carry opposite group velocities emulating the spin-up and spin-down states.

11.4 Acoustical Valley Hall Effect

The quantum states of energy extrema in momentum space describes the discrete valley degree of freedom [8]. This is known as the quantum valley Hall effect. There is rapid growing interest in this field due to its potential that can be used as a new form of information carrier such as spins in spintronics. The acoustic valley Hall effect is the transfer of the valley concept to classical wave systems. It has been shown that artificial phononic crystals can be used to engineer valley-like frequency dispersion [16]. Soon after this, the acoustic valley Hall effect and corresponding valley-protected edge states were theoretically predicted and observed experimentally. For instance, Lu et al. [17] first introduced valley states to acoustic waves using phononic crystals. The unit cells are hexagonal phononic crystals consist of triangular rods in a 2D waveguide with symmetries characterized by the rotation angle.

Lu et al. [18] also pointed out that the existence of a twofold Dirac degeneracy at the corners of the first Brillouin zone for any phononic crystal with $\alpha = n\pi/3$ is protected by the C_{3v} symmetry, whereas the degeneracy would be lifted for any other rod orientation breaking in the mirror symmetry.

11.5 Future Directions

Time-reversed symmetry plays an important role in topological acoustics. Using acoustic metamaterials in the form of phononic crystals, one is able to have quantum effects such as quantum Hall effect, quantum spin Hall effect, and quantum valley Hall effect built on a macroscopic classical structure. Topological acoustics is an example of the application of transport theory in condensed matter physics. Realizing topological robust and defect in sensitive wave guiding signal splitting and buffering could provide new directions for improved on-chip filters in mobile phones, surface acoustic wave sensors, enhanced coupling efficiency in touch screens and improve the robustness of bio-chemical sensing. Along this direction, efforts must be done in shrinking topological acoustical and mechanical properties in the micro-scale with the use of quantum phononic crystals [19] in order to be of relevance to the above-mentioned technologies.

References

1. Ginzburg,V.L., Landau, L.D.: On the theory of superconductivity. Zh. Eksp. Teor. Fiz. **20**, 1064 (1950). English translation in: L. D. Landau, Collected papers, p. 546. Pergamon Press, Oxford (1965)
2. von Klitzing, K., Dorda, G., Papper, M.: New method for high accuracy determination of the fine-structure constant based on quantized Hall resistance. Phy. Rev. Lett. **45**, 494 (1980)

3. Thouless, D.J., Kohmoto, M., Nightingale, M.P., Den Nijs, M.: Quantized Hall conductance in a two-dimensional periodic potential. Phy. Rev. Lett. **49**, 405 (1982)
4. Berry, M.V.: Quantal phase factors accompanying adiabatic changes. Proc. R. Soc. Lond. A. Math. Phys. Sci. **392**, 45–57 (1984)
5. Haldan, F.D.M.: Model for a quantum Hall effect without Landau levels: condensed-matter realization of the "parity anomaly." Phys. Rev. Lett. **61**, 2015 (1988)
6. Kane, C.L., Mele, E.J.: Quantum spin Hall effect in graphene. Phy. Rev. Lett. **95**, 226801 (2005)
7. Fu, L., Kane, C.L., Mele, E.J.: Topologicqal insulators in three dimension. Phys. Rev. Lett. **98**, 106803 (2007)
8. Ju, L., et al.: Topological valley transport at bilayer graphene domain walls. Nature **520**, 650 (2015)
9. Fleury, R., Sounas, D.L., Sieck, C.F., Haberman, M.R., Alù, A.: Sound isolation and giant linear nonreciprocity in a compact acoustic circulator. Science **343**, 516–519 (2014)
10. Ni, X., et al.: Topologically protected one-way edge mode in networks of acoustic resonators with circulating air flow. New J. Phys. **17**, 053016 (2015)
11. Zhu, Y., et al.: Experimental realization of acoustic Chern insulator, Preprint at https://arxiv.org/1801.07942,2018.
12. Souslov, A., van Zuiden, B.C., Bartolo, D., Vitelli, V.: Topological sound in active-liquid metamaterials. Nat. Phys. **13**, 1091 (2017)
13. Hassan, M.Z., Kane, C.L.: Colloquim: topological insulators. Rev. Mod. Phys. **82**, 3045 (2010)
14. Peng, Y.G., et al.: Experimental demonstration of anomalous floquet topological insulator for sound. Nat. Commun. **7**, 13368 (2016)
15. He, C., et al.: Acoustic topological insulator and robust one-way sound transport. Nat. Phys. **12**, 1124 (2016)
16. Zhang, X., Liu, Z.: External transmission and beating effect of acoustic waves in two-dimensional sonic crystals. Phy. Rev. Lett., **101**, 264303 (2008)
17. Lu, J., et al.: Observation of topological valley transport of sound in sonic crystals. Nat. Phys. **13**, 369–374 (2017)
18. Lu, J., et al.: Dirac cones in two-dimensional artificial crystals for classical waves. Phys. Rev. B. **89**, 134302 (2014)
19. Gan, W.S.: Quantum phononic crystals. In: Proceedings of ICSV24, paper no. 167 (2017)

Chapter 12
Application of Time-Reversal Acoustics to Signal Processing and Underwater Communications

12.1 Introduction

In this chapter, the signal processing aspect of time-reversal acoustics is considered. Time reversal is an iterative technique. In time-reversal acoustics, the time-reversal mirror (TRM) is used to transmit a plane wave towards the target and is reflected from it. The reflected wave returns to the TRM, where it appears as if the target has emitted a weaker signal. The TRM then reverses the signals and retransmitted the signal as usual and a more focused wave travels towards the target. The process is then repeated and this iterative procedure enables the sound waves to be more and more focused on the target.

From the signal processing aspect, time-reversal technique can be modelled as a matched filter. If one uses the delta function as the original signal, then the received signals at the TRM are the impulse response of the channel. TRM by sending the reversed form of the impulse responses back through the same channel, in effect autocorrelating it. The autocorrelation function has a maximum at the origin, where the original source was. In this case, the signal is concentrated in both space and time.

An advantage of time-reversal signal processing is that one does not need to have detailed information of the channel. By sending the sound wave through the inhomogeneous medium, one effectively measures it, as the retransmission just uses this data to focus the wave. Once the medium is reciprocal, one does not have to solve the wave equation to optimize the system.

Another useful property of time-reversal signal processing is the multipath propagation nature. Instead of compensating and correcting for multipath effects, multipath propagation can be used to their advantage by using the energy from all parts.

In the communication of information over noisy channels, the transmitted information code is subjected to a variety of signal distortions. This corrupts the fidelity of the information being transmitted. This reduces the effective capacity of the channel. Much research activities have been directed towards developing new techniques for correcting the effects of these disturbances. It is considered that time reversal is

W. S. Gan, *Time Reversal Acoustics*,
https://doi.org/10.1007/978-981-16-3235-8_12

useful methodology for this purpose. It can enhance signal levels in a reverberant environment with an array of transceivers.

Time-reversal signal processing can provide equalization which is the removal of channel distortion to extract the transmitted information. This is known as channel equalization which can provide greater channel capacity.

Most of the current methods for equalization require a priori knowledge of the channel with dominant characteristic of the channel. Time-reversal signal processing has the advantage that a priori knowledge of the channel is not needed for compensating for multipath distortion. The time-reversal receiver compensates for multipath distortions and achieves equalization by the use of an imbedded phase conjugation technique. The successful application of a time-reversal receiver can increase channel bandwidth and in the volume of information proportionally [1, 2].

12.2 Time-reversed Pocessing and Matched Filtering

From the signal processing aspect, time-reversal processing can be considered as an application of matched filtering [3] with a maximised output signal-to-noise ratio (SNR). There is the cross-correlation of the time-reversal replicant with the received signal to produce the optimal output [4]. The situation is more complex when dealing with spatiotemporal case. Here, the optimal matched filter is required to match the spatiotemporal channel medium impulse response which is Green's function besides matching the transmitted temporal function. Research shows that the spatial division multiple access (SDMA) depends on the spatiotemporal Green's function. Thus, time-reversal is the optimal implementation for SDMA [5]. Besides this, it has also been shown that time-reversal methods are also applicable to spatiotemporal phenomena satisfying an acoustic wave equation with time-reversal invariance property [6]. Time reversal represents the optimal spatiotemporal matched filter and maximizes the output SNR. It removes the aberrations created by an inhomogeneous channel. Besides this, time reversal has spatial diversity which is a potential application in signal communication.

The application of time reversal in signal communication started with the work of Parvulescu [1] in 1967 although it was published in 1995 due to security. First the Green's function of the ocean is estimated using a probe and then time reversed and retransmitted through the medium for focusing and eventually to achieve a high SNR gain. The formal theory of time reversal in underwater acoustics was developed by Jackson based on phase conjugation [2].

Underwater acoustic communication is regarded as one of the most challenging wireless communication due to the presence of large multipath propagation environment, large Doppler shift, fast time variant, and several channel fading. Time-reversal technique can achieve high data rate, reliable high frequency coherent communications, and reduces the system complexity compared with traditional methods. Time-reversal compresses dispersive acoustic energy both in time and space at the focal point. Also, this temporal compression recombines the dispersive multipath to reduce inter-symbol interference. Another useful property of time reversal is that the spatial focus of the time-reversal mirror mitigates the effect of channel fading. It can be divided into two main categories: active time reversal and passive time reversal.

12.3 Active Time-Reversal Communications

Active time-reversal communication [7] is communication from an array to a single point time reversal. The refracted and reflected waves can be converged to the focus point. This increases the signal level received at the focus and reduces the symbol interference. Cai et al. [7] proposed active time reversal in shallow water environments in combination with channel coding methods and adaptive channel equalization.

12.4 Passive Time-Reveral Communications

Unlike active time-reversal communication [7], passive phase conjugate communications are communicating from a single point to an array. Here, acoustic waves transmitted from a single point to an array. Here, acoustic waves transmitted from a single point are received at an array. During the process of time reversal, the received signals are demodulated which is equivalent to phase conjugation. Song et al. [8] found that there are two types of internal wave effects occurred during passive time-reversal communications. One is the long-term performance degradation in the scale of hours in the depressed waveguide when the internal waves passed through the acoustic track. The other is the rapid channel fluctuation giving rise to decreased channel coherence. Song et al. [9] presented the impact of spatial diversity in passive time-reversal communications between a vertical receiving array and a probe source by using a number of receivers distributed in space to take advantage of spatial diversity.

12.5 Challenging Issues facing Time-Reversal Communications

Time-reversal technique is a promising technique in underwater acoustic communication especially in shallow water environment due to its low complexity and robustness against inter-symbol interference due to multipath propagation. However, it faces several challenges as follows.

12.5.1 Receiver Design

Receiver design is one of the challenges in time-reversal underwater acoustic communication. Song et al. [10] proposed time-reversal multichannel receivers. Vector sensors were used for the receivers that showed significant size reduction. They used the vector sensor receiver for time reversal followed by a single-channel decision feedback equalizer to deal with the inter-symbol interference problem.

12.5.2 Application of Time-Reversal in other Advanced Wireless Communication Techniques

Candy et al. [11] proposed two time-reversal applications in multichannel underwater communications.

12.5.3 Interference Cancellation

Cho et al. [12] proposed a communication receiver architecture which used a time-reversal approach to suppress multiuser interference.

This is followed by a successive interactive interference cancellation scheme for cancelling the remaining interference.

12.5.4 Time-Reversal Underwater Acoustic Communication Experiment

A time-reversal underwater acoustic communication experiment [13] took place in May–June 2000 off the northern coast of Elba Island, Italy. In this experiment, an acoustic source/receiver array transmitted communication sequences to a vertical receiver array, 10 km away.

The probe source frequency was 3500 Hz and the focal depths were 60 and 80 m. Both time-reversed communication sequences and short 10-sec single source were transmitted. Time reversal is a two-way communication technique which first measures the channel response with a probe source. It then transmits the communication sequences. Time reversal can be considered as equivalent to the implementation of a matched filter of the impulse response of the waveguide. When a signal $s(t)$ is transmitted from a source in a waveguide, it is convolved with the impulse response of the waveguide $h(t)$, and the signal $r_i(t) = S(t)*\mathrm{h}(t)$ is then received on the ith element of the time-reversal mirror which is an array of N source/receiver transducers. The time-reversal mirror receives $r_i(t)$ and then retransmits the time-reversal version of the signal $r_i(-t)$.

The effect of time reversal is equivalent to the implementation of a linear matched filter yielding the autocorrelation of the impulse response of the waveguide. The signal received at the original source location $s_r(t)$ will be given as

$$s_r(t) = \sum_{i=1}^{N} s_i(-t)\star([(-t)\star i(t)]) \tag{12.1}$$

From (12.1), it shows that the quality of the time-reversal focus is improved with more TRM elements.

Time-reversal focus will be affected by the following factors: number of echoes, number of TRM elements, attenuation, noise, and temporal variation. Time-reversal resolution can be considered as linear broadband matched field processing since matched field processing is the computational implementation of time reversal.

The experiment conducted in June 2000 demonstrated that the time-reversal process recombined the temporal multipath resulting in reduced bit errors for communication.

References

1. Parvulescu, A.: Matched signal (MESS) processing by the ocean. J. A. S. A. **98**, 943–960 (1995)
2. Jackson, D.R., Dowling, D.R.: Phase conjugation in underwater acoustics. J. A. S. A. **89**, 171–181 (1991)
3. Hermand, J.P., Roderick, W.: Acoustic model-based matched filter processing for fading time-dispersive ocean channels: theory and experiment. IEEE J. Oceans Eng. **18**, 447–465 (1993)
4. Haykin, S.: Communication System. Wiley, New York (2001)
5. Paulraj, A.J., Papadias, C.B.: Space-time processing for wireless communictions. IEEE Signal Processing Mag. **14**(6), 49–83 (1997)
6. Fink, M.: Time reversal in acoustics. Contemp. Phys. **37**, 95–109 (1996)
7. Cai, L., Xu,W., Li, J., Gang, X.: Underwater acoustic MIMO communication based on active time reversal. In: Proceedings Prime Asia Conference, pp. 45–48 (2009)
8. Song, A., Badiey, M., Newhall, A.E., Lynch, J.F., DeFerrari, H.A., Katsnelson, B.G.: Passive time reversal acoustic communication through shallow water internal waves. IEEE J. Acoust. Eng. **35**(4), 756–765 (2010)
9. Song, H.C., Hodgkiss, W.S., Kuperman, W.A., Higley, W.J., Raghukumar, K., Akal, T., Stevenson, M.: Spatial diversity in passive time reversal communications. J. A. S. A. **120**(4), 2067–2076 (2006)
10. Song, A., Badiey, M., Hursky, P., Abdo, A.:Time revesal receivers for underwater acoustic communications using vector sensors. In: Proceedings IEEE Oceans Conference, pp. 1–10 (2008)
11. Candy, J.V., Chambers, D.H., Robins, C.L., Guidry, B.L., Poggio, A.J., Dowla, F., Hertzog, C.A.: Wideband multichannel time-reversed processing for acoustic communications in highly reverberant environments. J. A. S. A. **120**(2), 838–851 (2006)

12. Cho, S.E., Song, H.C., Hodgkiss, W.S.: Successive interference cancellation for time-reversal underwater acoustic channels. In: Proceedings IEEE Oceans Conference, pp. 1–7 (2009)
13. Edelmann, G.F., Hodgkiss, W.S., Kim, S., Kuperman, W.A., Song, H.C., Akal, T.: Proc. Oceans **2001**, 2231–2235 (2001)

Chapter 13
Application of Time-Reversal Acoustics to Seismic Exploration

13.1 Use of Time-Reversal Method for Location of Seismic Sources

In seismic exploration, the source location is essential. Unlike earthquakes, where the source can be located using triangulation, seismic source location is more challenging. The seismic signal exhibits no identifiable *P* and *S* wave arrivals. It emerges from the background noise without exhibiting sharp timing features. One needs a new method to trace such a signal back to its source. The technique of time reversal does not need sharp timing features and has great potential for such application. The time reversal is based on the time-reversal invariance of the sound wave equation which has no change in form when the time direction is reversed. That is, when t is replaced by $-t$. This is equivalent to that the observed wave propagation is symmetrical or the sound field has symmetrical property when energy dissipation is neglected. The positive time means the observation of the emission of sound waves from sources located beneath the earth surface and propagating towards sensors located on the earth surface. The negative time represents the sources of energy in sensors' positions and backpropagates through the rock mass to their sources. The energy is then focused at the point where the source is located. This can be generalized to superimposed events from a few sources. In this case, time-reversed propagation from sensors will be focused at the individual sources. The quality of the focusing depends largely on velocity structure of the geological media of the wave propagation. The following is an example of the procedure for the implementation of time-reversal method in geophysics. First, an unknown source emits a signal which is received at a set of seismic stations.

The seismic signals arrive at the different stations at different times are recorded. In the computer simulation, these signals are time reversed and rebroadcast from their respective stations into an elastic structure model of the earth. The earth model specifies the sound propagation velocity as a function of positions. The time-reversed wavefield in the earth model is then examined for focusing. This can reveal the nature and the location of the seismic source. In this computer simulation work, there are

W. S. Gan, *Time Reversal Acoustics*,
https://doi.org/10.1007/978-981-16-3235-8_13

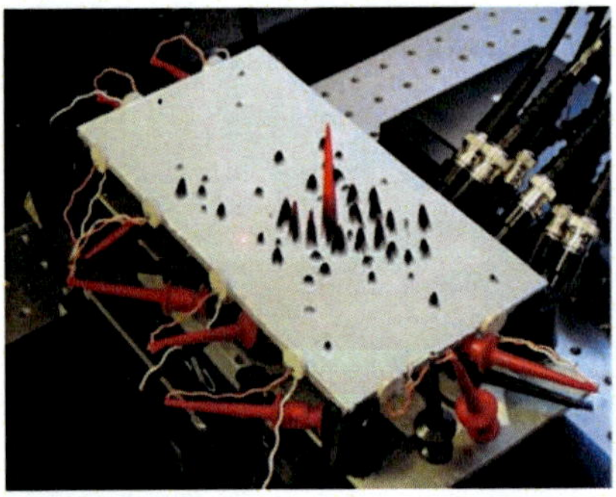

Fig. 13.1 Time reversal in the laboratory. A source on the underside of a thin aluminium plate, near its centre, emits a 20-μs strain pulse

three important ingredients: the velocity model, the set of seismic stations, and the choice of imaging fields for the interpretation of the results.

An example of the demonstrating of the application of time-reversal acoustics to seismic exploration in the laboratory is shown in Fig. 13.1.

The resulting displacements arriving at the plate's edges are recorded by eight detectors arrayed along the perimeter. The recorded signals are time reversed and rebroadcast from the detectors.

A laser vibrometre scans the plate's surface of the plate and maps the vertical component of its velocity due to vibrations created by the rebroadcast. The superposed peaks display the velocity map at the moment when the rebroadcast signals focus to reveal the original source (red peak). The lesser peaks are experimental artefacts. (After Larmat et al. 2014).

The idea of time-reversal imaging was first proposed by Parvulescu and Clay in 1965 for submarine communication [1]. The first application of the time-reversal method for source location was implemented in the early 1980s. Georg McMechan and colleagues at the University of Texas at Dallas [2] successfully imaged the source of a small 1983 aftershock earthquake in California by assembling all the ingredients of the time-reversal method. This is a purely acoustical rebroadcast was into a two-dimensional model of a 15-by-12 km vertical slice of the formation which contained the suspected source and the seismic stations. Their work revealed the limitations imposed by computing powers, the velocity model, and the seismic-station set.

13.2 Time-Reversal Seismic Source Imaging

Most of the work on the application of time-reversal imaging technique to seismic exploration has been for the case of single point source locations. Saenger et al. [3] have successfully applied time-reversal imaging to the location of multiple sources.

Anna Franczyk et al. [4] proposed time-reversal seismic source imaging using peak average power ratio(PAPR) parameter. It deals with sequences consisting of several events with the same origin time and with relatively small delays between them. A finite difference algorithm was used to model wave propagation in two dimension media. Based on the PAPR coefficient value, the location of source coordinates was performed in three steps. First, the exact locations of the source points and receivers and the synthetic seismograms for the given geological model were computed. This was followed by the time reversal of the full synthetic seismograms to perform the role of the source function. The final stage of the time-reversal rupture process imaging was the backward propagation of the seismic waves using the time-reversed synthetic seismograms. An iterative procedure of the backward propagation was performed with. The PARP coefficient was updated and stored for each time step of the propagation algorithm.

For the computation of the synthetic seismograms, the following acoustic wavefield equation was used:

$$\frac{\partial^2}{\partial t^2}p = \kappa\left[\frac{1}{\rho}\left(\frac{\partial^2}{\partial x^2}+\frac{\partial^2}{\partial z^2}\right)p + \nabla.f\right] \tag{13.1}$$

where $p(x, z)$ = acoustic pressure field, $\rho = \rho(x, z)$ = density of the medium, $\kappa(x, y)$ = bulk modulus, f = body force.

The above equation was transformed to a system of first order hyperbolic linear equations and was numerically using a staggered-grid, finite difference method which is stable for all values of a Poisson's ratio and characterized by relatively small grid anisotropy and small grid dispersion. In their work, Anna Franczyk et al. [4] used second-order temporal and fourth-order spatial operation under Cartesian coordinates. For the backward propagation part of the seismic wave, the same acoustic modelling as in the forward modelling is used. In the backward propagation, the time-reversed recordings are used as the sources of seismic waves. For the sound wavefield backpropagation, the same modelling scheme as the first stage computation is used. The time-reversed sound wavefield will focus in space and time in the computational node corresponding to the coordinates of the source point if the receiver configuration is sufficiently regular and dense. By visualization of the seismic field obtained during backpropagation of the inversed seismograms, the focusing point can be noticeable as an area of abnormal value.

13.3 The Imaging Parameters

The use of the maximum absolute pressure value (MAPV) as the imaging parameter and imaging condition shows to be a good indicator of the point source location. MAPV is defined as

$$\text{MAPV(x)} = \text{max/p(x, t)/} \tag{13.2}$$

$$T\ \varepsilon\ T$$

It has to be noted that the use of MAPV in time-reversed imaging can be a problem in the reconstruction of seismic events that are not separated in space or time [5]. Hence, Anna Franczyk et al. [4] proposed the application of the PAPR coefficient to improve the focus of seismic event detection. PAPR coefficient can be computed from the following coefficients:

$$\mathrm{PAPR}(x) = p_{\max}^2(x)/\left[p_{\mathrm{total}}^2(x)/T\right] \text{ and } p_{\mathrm{total}}^2(x) = \sum_{t=1}^{T} p^2(x,t) \quad (13.3)$$

where $p_{\max}^2$ = maximum energy of the signal recorded in the given computational node, p_{total}^2 = total energy in the given node, $p(x,t)$ = acoustic pressure value in each computational node, t = time index, T = number of computational time steps of the backpropagation of the seismic waves algorithm. It is found that a higher spatial resolution can improve the location of the seismic event sequences and can be obtained using time-reversal imaging with PAPR coefficient than with maximum values of a group parameter such as particle velocity or acoustic pressure or energy density or stress components.

13.4 Numerical Case Studies

Anna Franczyk et al. [4] performed numerical testing by using an inhomogeneous, layered model. They assumed that the P wave velocity increased from 3500 m/s in the first layer to 4200 m/s in the second layer and 4500 m/s in the third layer. The Ricker signal was used as a source function in the forward modelling. Their analysis was limited to the area in the vicinity of seismic sources where the focusing effect of the seismic waves was expected. The PAPR ratio and pressure changes were computed for every node of the computational grid during the second stage of modelling when the time-reversed seismograms were used as source function. The main difference between time-reversed imaging using the PAPR ratio and time-reversal imaging using MAPV is spatial resolution. The values achieved by the PAPR ratio in the source area are an order of magnitude greater than the values that it achieves outside this area. Values of MAPV outside the source area are reduced only a few times in relation to the values that they use for the source area. They also tested the advantage of time-reversal imaging using the PAPR ratio value in the detection of coordinates of seismic sources for the model with different excitation time of seismic sources.

Their numerical computation showed that time-reversal imaging using peak average power ratio (PAPR) parameter improved the spatial resolution in the detection of seismic sources location. The use of PAPR ratio as an imaging parameter improved the refocus for multiple sources with both the same excitation time and with a slight time lag between subsequent source emissions that simulate the rupture process. They showed results of the visualization of the backpropagation of seismograms reversed in time with the use of MAPV and PAPR ratio values.

References

1. Parvulescu, A., Clay, C.S.: Reproducibility of signal transmission in the ocean. Radio Electron. Eng. **29**(4), 223–228 (1965)
2. McMechan, G.A.: Geophys. J. R. Astron.Soc. **71**, 613–628(1982)
3. Saenger, E.H.: Time reverse characterization of sources in heterogeneous media. NDT E. Int. **44**(8), 751–759 (2011)
4. Franczyk, A., Leśniak, A., Gwiżdż, D.: Time reversal seismic source imaging using peak average power ratio (PAPR) parameter. Acta Geophysica **65**, 299–308 (2017)
5. Anderson, B.E., Griffa, M., Ulrich, T.J., Johnson, P.A.: Time reversal reconstruction of finite sized sources in elastic media. J. A. S. A. **130**(4), EL219–EL225 (2011)

Chapter 14
Application of Time-Reversal Acoustics to Ultrasound Non-destructive Testing

14.1 New Form of Ultrasound Scanning Mode

Ultrasound imaging rises to be a key method in non-destructive testing. The traditional method is the pulse echo technique in which a short pulse of ultrasound wave is emitted towards the sample and the reflected signals from various acoustic boundaries are recorded. There is also the transmission mode in which two ultrasound transducers, one as transmitter of signal and one as receiver of signal placed at opposite side of the sample to be inspected. There is a new third inspection mode known as time-reversal mode developed in Fink [1]'s laboratory. This new method increases the reliability and improves ultrasound focusing in particular for samples with complex shapes and enables a higher degree of certainty in detecting small defects. It is more difficult for ultrasound imaging in solids than in soft body tissue. Soft body tissue can only support longitudinal waves. Besides this, longitudinal wave velocity inside human body is relatively uniform over the frequency range used in medical diagnosis. This simplifies the pattern of echoes reflected back from the sample and enables a deeper penetration of sound wave and the exploration of the ultrasonic image in depth.

In ultrasound non-destructive testing, one is dealing with ultrasound imaging in solids which is a more complex situation than ultrasound imaging in human body tissues. Sound propagation in solids is more complex than in human body tissue. If the composition of the material varies or for a sample with complex geometry, defects detection will be even more difficult. Several transducers are needed to provide ultrasonic signals from a range of positions to detect small defects. This gives rise to the concept of phased array transducers. Here the ultrasonic arrays can be focused and steered electronically. The advantage lies in flexibility that a one-or two-dimensional transducer array can generate a focused beam at any angle and range and of any polarization. The focusing can be improved by connecting the transducer array to a set of electronic delay circuits that can compensate for the different propagation times between the focal point and each of the transducer elements. The interference pattern produced by the sound pulses can be used to focus and steer the beam during

W. S. Gan, *Time Reversal Acoustics*,
https://doi.org/10.1007/978-981-16-3235-8_14

the transmit mode. On the other hand, the delay lines ensure that the reflected waves recorded by each transducer are in phase before they can be summed and analysed during the receive mode. The delay lines deal with the divergence of the reflected waves from each scatterer in the material. This functions like a matched filter which simulates the action of a spherical transducer which produces a strong signal from a scatterer situated at its focal point. Thus, the focusing can be improved in both the receive mode and the transmit mode and produces high resolution. However, this technique has several limitations such as the exact interface geometry must be known a priori. Besides this, the transducers have to be positioned with extreme precision especially for the imaging of small defects. It also requires that the sound speed is constant in both the solid and in the liquid.

The time-reversed mode of ultrasonic imaging is based on the time-reversal symmetry property of the sound wave. The acoustic field equation is form invariant when t is replaced by—t and is valid if there is no heat dissipation. This means that the sound wave can backpropagate and retrace all the possible paths and converge at the original source as if time is going backwards.

14.2 Principle of Time-Reversal Imaging

The acoustic wave equation is given by the Helmholtz wave equation as:

$$\frac{1}{c^2}\frac{\partial^2}{\partial t^2}p = \nabla p \quad (14.1)$$

where p = acoustic pressure and c = sound velocity.

For the case of an isotropic solid immersed in a fluid, the acoustic field in the solid is given by

$$\rho\frac{\partial^2}{\partial t^2}u = (\lambda + 2\mu)\left[\text{grad}(\text{div}(u))\right] - \mu[\text{curl}(\text{curl}(u))] \quad (14.2)$$

where $\underline{u}$ = acoustic displacement, λ, μ = Lamė coefficients. Time-reversal invariance means that if $p(r, t)$ and $u(r, t)$ are solutions of (14.1) and (14.2), then $p(-r, t)$ and $u(r, -t)$ are also solutions of the same equations. Besides this, the $u(r, t)$ in the solid and the $p(r, t)$ in the fluid are linked by stress and strain relations at the fluid–solid interface that determines the unique solution for the acoustic pressure field $p(r, t)$ in the medium. This property can be used to achieve the optimal focusing of a pulsed wave on a point-like reflected location in the solid sample immersed in a fluid medium. Let $u(r, t)$ be the acoustic displacement in the solid and $p(r, t)$ the resulting acoustic pressure field in the fluid that propagates from a single point source located in a solid. The optimal way to focus on this source consists in a time reversal of the pressure field in the whole 3D volume, $p(r, -t)$ in the fluid and $h(r, -t)$ in the solid.

One can reduce the time-reversal operation from a 3D volume to a 2D closed surface resulting in the concept of a closed time-reversal cavity by using Huygen's principle.

14.3 The Time-Reversal Mirror

To perform an ideal experiment on time reversal, the source of the sound wave has to be located inside the solid. This sound source emits a short acoustic pulse consisting of both longitudinal waves and shear waves with some of the acoustic energy transmitting through the solid/liquid interface. After the transmission, the transverse wave will be converted into the longitudinal wave. The resulting acoustic pressure field is measured by transducers on a closed surface completely surrounding the fluid. This is known as the time-reversal cavity.

After the disappearance of the wave, the signals are time reversed and reemitted into the field and travel towards the initial source with mode conversion taking place in the reverse direction. This generates both transverse waves and longitudinal waves in the solid which converge at exactly the same time at the source. This time-reversed technique works well even for complex interfaces for a relatively small source. However, some of the information will be lost due to the fact that the horizontally polarized shear waves contained in the pulse radiated from the source cannot be converted into longitudinal waves and cannot be time reversed. Besides this, the time-reversed field encounters diffraction with the diffraction spot similar in size to the smallest wavelength in the pulse. Sources smaller than this cannot be resolved in full. Also the time-reversal cavity is unable to sense the evanescent waves which contain information on smaller length scales.

In practice, a time-reversed cavity is extremely difficult to realize because time-reversal cannot be obtained on a surface which completely surrounds the fluid but only a limited area. This gives rise to the concept of time-reversal mirror. The time-reversal mirror consists of an array of transducers. Each transducer comes with its own electronics: digital memory, A/D converter, detection amplifier, and a programmable generator to synthesize the temporally inverted signal stored in the memory. Unlike ordinary mirror which produces virtual images, the time-reversal mirror produces a real acoustic image at the position of the initial source. The time-reversal technique is adaptive. It is capable of focusing on defects beneath curved surfaces. The time-reversal mirror allows the conversion of a divergent wave emitting from a defect into a convergent wave focusing on it. The imaging can be achieved in real-time and with simple signal processing. It is a natural focusing process matched to the defect shape and to the geometry of the liquid–solid interface and able to compensate the distortion induced by complex liquid–solid interfaces.

Another advantage of time-reversal technique is the speckle noise reduction capability. If the speckle noise arises from a random microstructure with scale less than the wavelength, then the time-reversal process cannot exactly refocus on the speckle noise sources. This will give rise to each individual scatterer giving a time-reversed wave which focuses back on it with a diffraction spot. The interface between these diffraction spots produces a complex pattern which does not match the exact scatterer distribution. This property enables one to develop processing technique to distinguish high-level incoherent speckle noise from low-contrast flaw during the inspection.

14.4 Experiment on Application of Time-Reversal Mirror to Ultrasound Nondestructive Testing

As a practical example of the application of the time-reversal technique to ultrasound non-destructive testing, the following example is quoted from the work of the Matthias Fink group [2]. They have used a 128 channels time-reversal mirror to detect low-contrast defects within the titanium alloys used in jet engines. The central ultrasound frequency used is 5 or 7.5 MHz and the lateral dimension of the time-reversal mirror built with piezocomposite material is between 60 and 120 mm. To perform the experiment, the time-reversal mirror transducer array and the titanium sample with different defects are immersed in water facing each other. The central part of the transducer array is used to illuminate the titanium sample with the echoes recorded by the whole array. During the experiment, the incident sound beam is converted only into the longitudinal wave for propagation in the titanium sample because the speckle noise is weaker for longitudinal waves than for shear waves in titanium.

The signals from a region containing low-contrast defect are shown in Fig. 14.1 The vertical axis represents the code number of the array element while the horizontal line represents the sound echo amplitude received by one element as a function of time or reflecting target depth. Fig.14.1 shows the echoes from the two titanium/water interfaces. Between the two echoes, the speckle noise caused by the titanium microstructure is shown. However, the echo from the defect is hidden by the high level of speckle noise. With a short time window, a portion of the acoustic signal

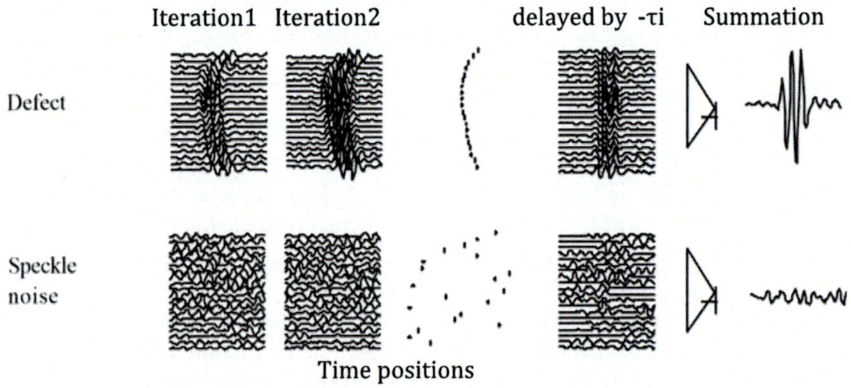

Fig. 14.1 The signals from a region containing low-contrast defect. (After Fink et al.)

is selected, time reversed and reemitted from the whole array. The new recorded data denoted by iteration shows a well-contrasted wavefront appearing within the window range. The sound wave backscattered by the defect that intercepts the 2D array with oblique incidence gives rise to wavefront phase modulation. A similar wavefront is produced by a second iteration of the time reversal.

Repeating the experiment for a zone without defect, a different result is obtained (Fig. 14.1). It shows two different noise echo patterns originating from the microstructure after each illumination due to different illuminated beams.

This time-reversal technique reduces to self-focusing only in the transient mode. By focusing in the receive mode, a better signal-to-noise ratio can be obtained. For the receive mode, before summation, one needs to compensate the different time delays between the defect and the array elements. The estimation of the delays is required only if the window contains a defect. One has to be careful when estimating delays in a window containing only speckle noise.

References

1. Chackroun, N., Fink, M., Wu, F.: Time-reversal processing in nondestructive testing. IEEE Trans. Ultrason. Ferroelec. Freq. Contr. **42**(6), 1087–1098 (1995)
2. Fink, M., Prada, C., Cassereau, D., Kerbrat, E.: Time reversal techniques in nondestructive testing

Chapter 15
Application of Time-Reversal Acoustics to Medical Ultrasound Imaging

15.1 Application of Time-Reversal Acoustics to Focusing System for Drug Delivery

The application of therapeutic ultrasound requires focusing of acoustic energy to a specific region of tissue which is the therapeutic target while minimizing the exposure of healthy tissue surrounding the target. It will be a great challenge if the target is deep inside the body. The conventional methods of focusing in medical ultrasound depend on lenses and concave transducers, and phased array transducers and these methods work well for homogeneous media but not for heterogeneous media. This is due to the multiple scattering and diffraction of sound wave and wavefront distortions in a heterogeneous medium. The phased array transducers with their multi-elements are more flexible to handle the heterogeneity due to the elements being fed from individual power amplifiers and phases and amplitudes can be varied at these elements to enable the control of the phase relations over the entire array's aperture to form a wavefront of the shape required to correct the aberration caused by the medium's heterogeneity. However, the phased array transducers have complicated, intricate design, and do not provide the required improvement and also generate undesired grating lobes that focus acoustic energy outside the targeted region.

In view of the above problem, the time-reversal acoustics is an effective method of focusing acoustic energy deep inside the heterogeneous and multiple scattering media such as biological tissues. Time-reversed acoustics-based ultrasound focusing systems have been developed for focusing ultrasound wave onto specific locaions in heterogeneous media in several physical and biological applications [1, 2].

To deliver drugs into the brain, convection-enhanced delivery is used. It infuses drugs directly into the brain interstitium. William Olbricht et al. [3] proposed the combination of convection-enhanced delivery with time-reversal acoustics in a focusing system that uses a smart needle to simultaneously infuse fluid into the brain and to provide the necessary feedback for focusing ultrasound using time-reversal acoustics.

W. S. Gan, *Time Reversal Acoustics*,
https://doi.org/10.1007/978-981-16-3235-8_15

Convection-enhanced delivery (CED) is used for drug delivery to the brain for treatment of several neurological disorders. The blood–brain barrier which blocks the systemic drug delivery is circumvented by CED which uses a small needle to insert into the brain skull and infuse drugs through a small hole in the skull. Infused liquid containing therapeutic compounds flows radially outwards from the needle tip into the surrounding brain tissue. The dominant interstitial transport mechanism near the needle tip is convection. So far the results of CED therapy are not promising due to the inadequate distribution of infused drugs into the brain parenchyma.

Ultrasound-enhanced drug delivery can lead to enhanced mass transfer. Ultrasound enhances both convection and diffusion through neural tissue [4]. The underlying mechanisms of ultrasound-enhanced drug delivery that lead to enhanced mass transfer are acoustic cavitation and resulting microstreaming [5, 6]. The unfocused ultrasound used in drug delivery is targeting at a relatively large volume of tissue such as the entire hemisphere of the brain. The time-reversal acoustics method combined with CED can enhance the mass transfer of infused compound by selectively exposing a much smaller portion of the brain to focused ultrasound.

To use the time-reversal acoustics method to focus acoustic energy, it is necessary to place a beacon in the medium that provides the initial signal from the target region. William Olbricht et al. [3]used a proprietary smart needle to infuse fluid into the brain. The tip of the needle is equipped with a miniature piezotransducer to accurately focus ultrasound precisely at the point where fluid enters the tissue. This transducer provides the required feedback signal for the time-reversal acoustics relationship between the target site in the brain and the transmitter. Hence, this smart needle besides delivers fluid to the tissue also provides the signal for the time-reversal acoustics focusing at the delivery site.

15.2 Experimental Procedure for Time-Reversal Acoustics with the Convection-Enhanced Delivery(CED) for Drug Delivery

Their smart needle is an infusion catheter, custom designed, and contains a small hydrophone. It has a 30-gauge stainless steel needle equipped with a longitudinal resonating piezotransducer with centre frequency of 0.5 MHz. The transducer is attached to the distal end of the needle. It is equipped with a wire conducting the recorded signal to the electrical connections in an aluminium guide to be holding the needle. The smart needle hydrophone has to be calibrated with a low-frequency hydrophone in a water beaker.

During the CED infusions, the smart needle records the acoustic signal at the point where the fluid leaves the needle and enters the brain. This recorded signal is then time reversed, synchronized, amplified, and applied simultaneously to all the transmitters. The modified signal then focuses accurately at the target, a known volume of tissue surrounding the needle tip.

The experiments were performed on rodents. The results of preliminary experiments conducted with brain-mimicking phantoms were used for choosing the ultrasound parameters for the experiment. The brain mimic used for the CED experiment is a small, clear plastic vial filled with 0.6% agarose gel. An ultrasound frequency of 750 kHz was used with a time-averaged intensity of about 20mW/cm^2.and peak-to-peak acoustic pressure of 150–250 kPa. The smart needle was then held in a three-dimensional positioning system and lowered into the gel. The time-reversal acoustics system was activated once the smart needle was in place. 10 μl of water containing red food colouring was infused through the needle into the gel in about 1 min. Ultrasound exposure was continued for 30 min following the infusion of the dye into the gel.

After the ultrasound exposure, the gel was removed from the plastic vial and cut in half in a plane containing the needle insertion point. Each half of the gel was analysed with MATLAB software and photographed. With the coloured part of the image separated from background, the boundary of the dye distribution was circumscribed, the diameter of the coloured area, close to circular in every case was measured. The volume of distribution of dye was calculated for a spherical distribution based on the measurement.

Then the infusions into anaesthetized rodents were performed. Figure 15.1 showed a schematic of the rodent experiment.

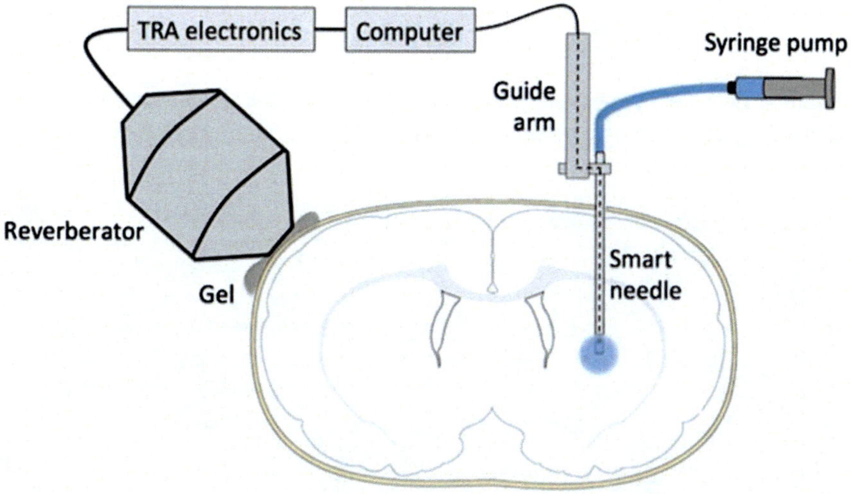

Fig. 15.1 (Colour online) A schematic of the rodent experiment. The smart needle is attached to a guide arm that allows precise placement of the needle. A syringe pump infuses tracer through the smart needle. A hydrophone is freely suspended at the distal end of the smart needle to measure the acoustic signal there. A wire from the hydrophone is threaded up the needle and guide arm and connected to a computer that records the acoustic signal. The resulting signal is time reversed, and the time-reversed signal is sent to an external reverberator coupled with gel to the outside of the rodent skull. (After William Olbricht et al. 2013)

The focal volume was about 2 mm in diameter. Ultrasound parameter was set to 1 MHz frequency with 8 μs pulse duration and burst mode with eight pulse repetitions.

Time-reversal acoustics focusing and ultrasound exposure were initiated at the start of the experiment and maintained for 30 min. The animal was euthanized by pentobarbital injection immediately after the experiment. The brain was removed and stored in a 4% paraformaldehyde for 24 h, then in a 30% aqueous sucrose solution for 48 h and in a 60% aqueous sucrose solution for 48 h. Then the brain was frozen on dry ice in optimal cutting temperature compound and subsequently sliced in a 50 μm slices on a microtome.

Digital image files were cropped to include the rodent brain's caudate putamen with a locked aspect ratio. The disturbance volume of infused dye for each experiment was calculated after processing the modified images with MATLAB. The ratio V_d/V_l can be calculated with the known volume of fluid infused V_i and the measured distribution volume V_d, the ratio V_d/V_l can be calculated. Since infused fluid flows only through the interstitial space between cells, this ratio is typically larger than unity.

15.3 Experimental Results for Combining Time-Reversal Acoustics with the Convection-Enhanced Delivery (CED) for Drug Delivery

There is time-reversal acoustics (TRA) focused ultrasound enhancement in brain-mimicking gels. This can be shown in Fig. 15.2

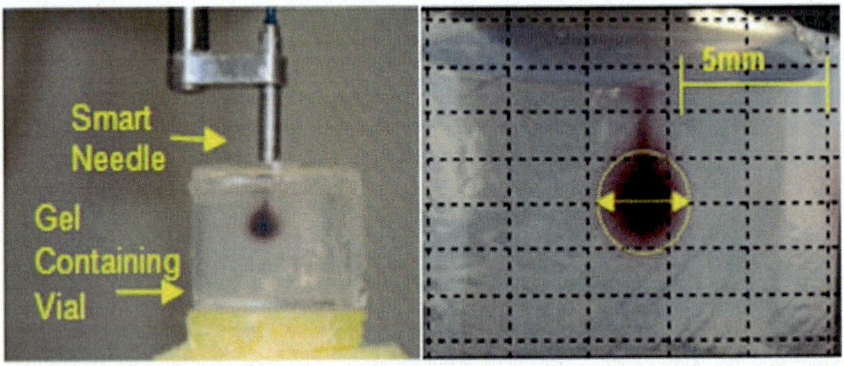

Fig. 15.2 (Colour online) Infusion of dye into an agarose gel. Left: The smart needle containing the TRA hydrophone is inserted into 0.6% agarose gel contained in a transparent vial. Right: The infused dye at the completion of the experiment. The infusion cloud is nearly spherical with diffuse edges. The yellow circle shows the boundary of the infusion cloud. (After William Olbricht et al. 2013)

This shows the typical infusion into agarose gel at the end of the experiment after the removal of the smart needle as a near spherical distribution of the infusate with a diffuse boundary. The meaning of distribution volume means the volume of the gel coloured by the infused dye at the end of the experiment. The distribution volume without ultrasound was compared with the distribution with ultrasound to determine the effect of TRA-focused ultrasound on the CED infusion. It was found that the distribution volume with ultrasound was larger than the distribution volume without ultrasound by a factor of 7.2 ± 0.4 for the case of infusion into pure gels.

To find the effects of time-reversal acoustics focused ultrasound on the distribution of infused tracers, experiments were carried out in anesthetized rodents. A small volume of tracer was introduced into the caudate, a grey matter for the measurement of the spread of the enhancement in distribution volume and the spread of the infusate due to time-reversal acoustics focused ultrasound without the infusate spreading beyond the caudate into the surrounding heterogeneous tissue medium. The results of the time-reversal acoustics focused ultrasound enhancement in anaesthesized rodents was shown in Fig. 15.3.

It shows the various brain slices carried out in various experiments. All the slices shown in the figure contained the smart needle. Figure 15.3a and e showed the smaller areas of dye coverage pertaining to infusion of Evan's blue dye(EBD) and albumin-conjugated (EDB), respectively, without ultrasound and microbubbles. Figure 15.3b for EBD and Fig. 15.3f for albumin-conjugated EBD showed the measureable enhancement of the distribution when time-reversal acoustics focused ultrasound was applied. There was also an increase in the infusate as shown in Fig. 15.3c for EBD and Fig. 15.3g for albumin-conjugated EBD. When the infusate containing time-reversal acoustics focused ultrasound and stabilized microbubbles was applied, the greatest enhancement of dye coverage occurred as shown in Fig. 15.3d for EBD and Fig. 15.3h for albumin-conjugated EBD.

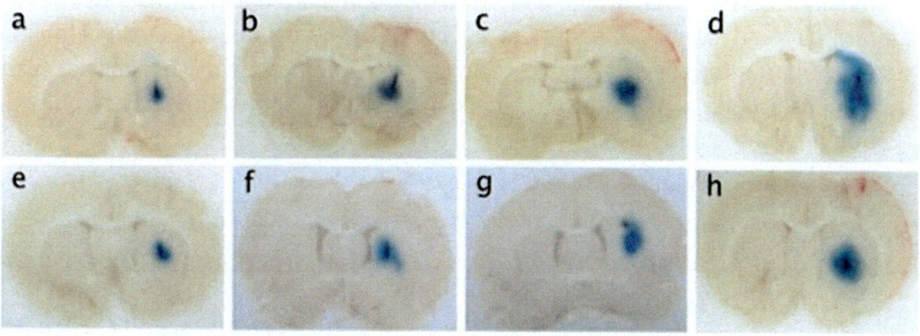

Fig. 15.3 (Colour online) Typical coronal slices of rat brains illustrating the infusion of EBD and albumin-conjugated EBD. **a** EBD, no ultrasound, no microbubbles; **b** EBD, ultrasound, no microbubbles; **c** EBD, no ultrasound, microbubbles; **d** EBD, ultrasound, microbubbles; **e** albumin-conjugated EBD, no ultrasound, no microbubbles; **f** albumin-conjugated EBD, ultrasound, no microbubbles; **g** albumin-conjugated EBD, no ultrasound, microbubbles; **h** albumin-conjugated EBD, ultrasound, microbubbles. (After William Olbricht et al. 2013)

The experiments in live, anaesthesized rodents showed the use of a smart needle to infuse fluid in CED coupled with time-reversed acoustics focused ultrasound in the tissue surrounding the tip of the needle improved the volume of distribution of EBD and albumin-conjugated CED. With the infusate seeded with stabilized microbubbles, the enhancement in tracer penetration can be improved. Postmortem detailed analysis of brain slices showed that the shape of the distribution volume is close to spherical in the tissue region immediately next to the needle tip with the shape more elongated in the anterior–posterior direction away from the needle tip. It is noted that the effect of the time-reversal acoustics focused ultrasound on tracer penetration propagated beyond the focal volume of the ultrasound. This is beneficial to CED which is targeting at promoting the penetration of therapeutic molecules dissolved in the infusate.

15.4 Medical Application of the Nonlinear Effects of Time-Reversal Acoustics

Time-reversal acoustics focusing helps in the focusing of ultrasound wave in heterogeneous medium. Pervulescu and Clay [7] were the first to apply the principle of time-reversal invariance in physics. They tested this in underwater acoustics in 1965 in the ocean by using a single element receiver/transmitter pair at a distance of 20 miles. Time-reversal acoustics focusing is very useful for the precise focusing of acoustic energy to chosen parts of the human body for medical application. In this section, the extraction of the nonlinear effects of time-reversal acoustics will be considered. The usual multiple scattering and numerous reflections from boundaries that destroy focusing in a common system led to the enhancement of focusing ability instead in a time-reversal acoustics focusing system.

Examples of the medical applications of the nonlinear acoustic effects are for bone assessments and earlier osteoporosiss detection [8] and nonlinear acoustical imaging of ultrasonic contrast agents [9]. It is of interest to combine the time-reversal acoustics method with nonlinear acoustics method for medical applications. The following section provides examples of this.

15.4.1 Application to Blood Pressure Measurement in Heat Chamber

The procedure is to inject ultrasound contrast agents (UCA) into the blood and the scattering from the UCA in the heart is used for time-reversed acoustics focusing. The level of the second harmonic signal scattered by the UCA can be used for the determination of the ambient pressure based on the measurements of the second harmonics. The preliminary experiments to demonstrate the dependence of the second harmonics scattered by the UCA on the ambient pressure were conducted [10]. The experimental results were in good agreement with the theoretical model describing the dependence of the second harmonics on the ambient pressure given by Zabolotskaya and Soluyan [11].

15.4.2 Application to Assessment of the Human Calcaneus

Alexander Sutin et al. [10] conducted preliminary experiments using nonlinear time-reversal acoustics to assess the human calcaneus. This was based on the concentration of the acoustic energy of two signals with different frequencies in a small volume of bone. The generated signal with a sum frequency was measured. A needle hydrophone was used to detect the acoustic signal. The purpose of the sum frequency measurements was to separate the nonlinear effects produced by bones from the nonlinear effects generated by electronics and transducers that led to second harmonics generation. A parameter relating the amplitude of the sum frequency to the amplitude of the second harmonics of the initial signals was introduced.

References

1. Fink, M., Montaldo, G., Tanter, M.: Time-reversl acoustics in biomedical engineering. Ann. Res. Biomed. Eng **5**, 465–497 (2003)
2. Pernot, M., Aubry, J.F., Tanter, M., Boch, A.L., Marquet, T., Kujas, M., Seilhean, D., Fink, M.: In vivo transcranial brain surgery with an ultrasonic time-reversal mirror. J. Neurosurg. **106**, 1061–1066 (2007)
3. Olbricht, W., Sistla, M., Ghandi, G., Lewis, G., Jr., Sarvazyan, A.: Time-reversed acoustics and ultrasound-assisted convection enhanced drug delivery to the brain. J. A. S. A. **134**(2), 1569–1575 (2013)
4. Lewis, G.K., Jr., Olbricht, U.L., Lewis, G.K.: Acoustically enhanced evanescent blue dye perfusion in neurological tissues. J. A. S. A. **2**, 20001–200017 (2008)
5. Pitt, W.G., Husseini, G.A., Staples, B.J.: Ultrasonic drug delivery-a general review. Expert Opin.Drug Deliv. **1**, 37–56 (2004)
6. Zarnitsyn, V.G., Kamaev, P.P., Prausnitz, M.R.: Ultrasound-enhanced chemotherapy and gene delivery for glioma cells. Technol. Cancer Res. Treat. **6**, 433–442 (2007)
7. Parvulescu, A., Clay, C.S.: Reproductibility of signal transmission in the ocean. Radio Electron. Eng. **29**(4), 223–228 (1965)
8. Muller, M., Sutin, A., Guyer, R., Talmant, M., Laugier, P., Johnson, P.A.: Nonlinear resonant ultrasound spectroscopy (NRUS) applied to damage assessment in bone. J. A. S. A. **118**(6), 3946–3952 (2005)
9. Schrope, B., Newhouse, V., Uhlendorf, V.: Simulated capillary bloodflow measurement using a nonlinear ultrasonic contrast agent. J. A. S. A. **90**(4), 2358 (1991)
10. Sutin, A., Salloum, H.: Prospective medical application of nonlinear time reversal acoustics. In: Proceedings of 21st International Symposium on Nonlinear Acoustics, vol. 34, 020003 (2018)
11. Zabolotskaya, E.A., Soluyan, S.I.: Emission of harmonic and combination frequency waves by air bubbles. Soviet Phys. Acoustics **18**, 396–398 (1972)